Thylacine

The History, Ecology and Loss of the Tasmanian Tiger

Editors: Branden Holmes and Gareth Linnard

Reprinted 2023.

A catalogue record for this book is available from the National Library of Australia.

ISBN: 9781486315536 (pbk)
ISBN: 9781486315543 (epdf)
ISBN: 9781486315550 (epub)

How to cite:
Holmes B, Linnard G (Eds) (2023) *Thylacine: The History, Ecology and Loss of the Tasmanian Tiger*. CSIRO Publishing, Melbourne.

Published by:

CSIRO Publishing
36 Gardiner Road, Clayton VIC 3168
Private Bag 10, Clayton South VIC 3169
Australia

Telephone: +61 3 9545 8400
Email: publishing.sales@csiro.au
Website: www.publish.csiro.au|
Sign up to our email alerts: publish.csiro.au/earlyalert

Front cover: Thylacine at Beaumaris Zoo, 1936 (photo by Ben Sheppard)

Edited by Kerry Brown
Cover design by Cath Pirret
Typeset by Envisage Information Technology
Printed by Ingram Lightning Source

CSIRO acknowledges the Traditional Owners of the lands that we live and work on across Australia and pays its respect to Elders past and present. CSIRO recognises that Aboriginal and Torres Strait Islander peoples have made and will continue to make extraordinary contributions to all aspects of Australian life including culture, economy and science. CSIRO is committed to reconciliation and demonstrating respect for Indigenous knowledge and science. The use of Western science in this publication should not be interpreted as diminishing the knowledge of plants, animals and environment from Indigenous ecological knowledge systems.

The authors are generously allocating all their royalties from the sale of this book to support research into Devil Facial Tumour Disease, a devastating and contagious disease now threatening the wild population of Tasmanian devil – the world's current largest marsupial carnivore.

Apr23_RP_ILS

Foreword

Nick Mooney

In 1986, I was taken to the site of a thylacine capture of 65 years before. I felt a chill as an aged but animated Peter, a teenager when he and his father caught the animal, found the now decrepit log exactly where he had described. There among the crumbling hardwood with its fungi, moss and lichen, were embedded green strands of a crusted copper 'necker' snare. I stood aghast taking in his excited reconstruction, then felt robbed.

I'm saddened to know I will never see a thylacine. Our bush is much poorer for its apparent absence and we should never forget our recklessness. If there are none now, perhaps there have been none for a long time as über-sceptics have insisted. I am bemused by the human contradiction in wanting to put things to bed, yet knowing one cannot prove a negative. The capture of the *last known* wild thylacine in 1933 was routinely touted as the capture of the *last* thylacine and the death of the *last known* in 1936 as the death of the *last*, clearly homocentric hubris. Sorry, but I have trouble swallowing that chain of assumptions.

What does the continued enthusiasm for *Thylacinus* tell us? The disciplines presented in this book hopefully reflect a new consciousness regarding the thylacine, so evocatively and misleadingly called the Tasmanian tiger. It seems that many people are now aware of what has been lost with the demise of this largest of recent carnivorous marsupials. Certainly, the sudden appearance of lethal disease driving a crash in numbers of Tasmanian devils reminds us of the unexpected changes that can happen on our watch.

Recent history of the thylacine was not bizarre. In fact, it was a grimly normal anthropogenic process – overhunting of a closed population leading to serial collapse of local populations. Other factors may have played out in the diminished population. Perhaps the species was unusually susceptible to habitat change such as wrought by colonialists and at the end fertility and immune systems may have degenerated. Similar situations are playing out in countless theatres on land, sea, and air around the world as the 'sixth extinction'. In Tasmania it was government sanctioned, sheep farmer greed that drove this debacle, in retrospect corrupt. How ironic that museums and zoos, so integrated with conservation now, may have been a final nail in the thylacine's coffin.

Thylacines did just fine for thousands of years in Tasmania before the British invaded and had there been just a smidgen of sobriety in colonial governance then this intriguing and ecologically important animal would still be with us. As it stands, wedge-tailed eagles are possibly Tasmania's most important terrestrial animal predator and the last apex predator, devils being a relatively 'blunt instrument'. It is one of the reasons I have worked long and hard on eagle conservation. In fact, the preferred diet and habitat of wedge-tailed eagles and what we know of thylacines seem to overlap so much that I use eagles as a proxy when considering potential thylacine distribution. I am certainly not the first European Tasmanian to notice similarities – the Buckland and Spring Bay Tiger and Eagle Extermination Association of 1884 did just that and stands testament to colonial predator hysteria.

There seems a new enthusiasm to figure out the myriad missing pieces (huge chunks really) in our knowledge of thylacine behaviour and ecology. Much of this must be argued and modelled and that, of course, results in descriptions with varying probabilities of accuracy. The result can sometimes conflict with historical observations but regardless, it piques our interest. If nothing else, these varied efforts surely drive home the advantages of studying animals while they are common. Paradoxically, funding is usually only available for the endangered. Other parallel efforts continue to try and prove the likelihood of extinction wrong but in this the enthusiastic amateurs have done no better than the consummate professionals. Perhaps the declining state of the natural world prompts people to clutch at straws. On the other hand, it may reflect hope and determination to rescue nature, as Sir David Attenborough and so many young people so firmly demonstrate. It may also reflect the easy availability of technology such as trail cameras and social media's ability to generate fashions via 'click bait'. It may (and probably should) also reflect collective guilt and simple curiosity.

There are arguments that thylacines would still be with us if we had domesticated them, but I beg to differ. Tasmanian aboriginals lived among these animals for many thousands of years and there is no observation they used them for anything except food. These resourceful people did, however, quickly adopt dogs on their introduction and in doing so clearly voted on comparative use.

Decades ago, there seemed a real chance that something amazing might happen as rediscovery, but searches were ultimately hampered by rudimentary automatic cameras and no recorded vocalisations for use in surveys. There were also no reference scats and DNA analysis was very difficult, so Tasmanian Parks and Wildlife Service (with its beautiful acronym TASPAWS) developed tests on bile salts to try and identify scats by a process of elimination, but the test was not specific enough. Yet authorities hedged their bets and it remains illegal to even try and catch one – yes, someone was successfully prosecuted.

While recently re-examining the footprints cast by the indomitable Trooper Fleming with Michael Sharland at the Jane River in 1938, I thought of their practical approach to the issue on the shoulders of the Great Depression and war. That search was explorative and rather matter-of-fact. They seemed convinced thylacines still existed but in peril.

Subsequent searchers attempted capture but were, I think, very unlucky, especially since they could call on many people who had caught thylacines. David Fleay even claimed that one got away. It was the mid-1940s and he may well have been right. What a pity that his modern conservation ethic had not come a couple of decades earlier. The irascible Eric Guiler with Inspector Hanlon later carried out prolonged trapping searches at excellent sites but to no avail. There have been other credible searches, but the efficacy of most has been wildly overstated. Some searches have been little more than headlines. Now we have fresh minds applied to the issue and a variety of effective new search and analysis techniques.

More recently, others have turned mathematics to the task and produced papers claiming astronomical odds against thylacines existing and yet others using an optimistic interpretation of much the same material argue that thylacines may just still exist. Such models make foundational assumptions about detectability and that is what drags them back to the challenge of coinciding a sure detection method with a few animals across ~60 000 km^2. Simply put, once a species drops below a certain landscape density, detection methods progressively collapse in reliability until the point where luck takes over from systematic method. But probability is relentless and great persistence with a logical detection method(s) can find something and if not, come as close to proving a negative as you can. We are still a distance from that point I think, but the gap is small and closing.

Trail cameras have breathed new life into searching for all manner of animals. At any one moment I calculate there are at least 500 camera traps set by private wildlife enthusiasts, consultants and university or government researchers in Tasmania. These cameras are in national parks, on crown land and state forest, on farms and mine sites, near roads, in wilderness on beaches and mountains, in caves and back yards; pretty well anywhere in Tasmania and undoubtedly 'in a place near you'.

Many cameras that I know of are in perfectly sensible places to 'catch' thylacines if they were extant, some in logically excellent places with exactly that in mind. More than one searcher has a packet of cameras relaying pictures back to their office or home and I have great fun helping (in confidence) identify mysterious pictures for them. Indeed, I sit here now with my friend Jason Wiersma monitoring phones for images of wedge-tailed eagles sent by SMS from cameras he has deployed for research.

For all their convenience, even trail cameras are not foolproof however. On any day about one-third will be ineffective either through mechanical or electrical failure (repeatedly heating and cooling can fatigue the solders), condensation or drowning, getting knocked over by animals (I recently had a wedge-tailed eagle walk up and 'foot' one, by chance opening it and spilling the batteries), flat batteries, full cards, uncharged SIM cards, being stolen or, worst of all, not being switched on. Regardless, they are a lot better than the clunky SLRs (the camera not the military rifle!) or the super 8 movie cameras triggered by pressure mats (surprisingly effective when not chewed by devils) that we used to have. Essentially unlimited numbers of discrete cameras are exactly what we dreamed of decades ago in those heady days in north-west Tasmania's Arthur River forests. The mission was to try and confirm the 1982 thylacine sighting report at close-quarters by my then colleague Hans Naarding, a report Hans has never wavered from.

Images still need curating of course and the usual poor provenance of claimed evidence seriously compromises its use whatever the image shows. Too often the first views we get now are already tinkered with and/or enmeshing in Facebook. Fortunately, digital forensics is now an advanced science and images can be deeply scrutinised if need be.

Some people are carefully resurrecting old skills of tracking and then applying digital technology to help interpret footprints. Drones are common and email has allowed the quick marshalling of a variety of opinions on possible evidence. DNA analysis of scats is routine. The modern library of all known images is a fantastic resource allowing us to compare reports and images with what the animal can look like with its variety of body conditions and postures through different lighting. So, the means are there but what about the ends?

Beyond the lack of irrefutable evidence, I regret to say there are two big problems with the proposition that thylacines still exist.

Firstly, too many 'unlikely' ducks have to be lined up. Thylacines were not exceptionally wary of people and as Europeans infiltrated Tasmania were not all that hard to find. Thousands were killed often by crude means. They were guided along fences into pit-falls, dogs caught them and some were shot. A few were even clubbed to death when cornered near people. Many were caught in snares set for wallaby, both unwary species that did not require care with human scent etc. As thylacines became rarer it is possible the more wary predominated, but remove that pressure and animals quickly revert; consider the amazing scenes of animals occupying towns during COVID-19 lockdowns. People often use descriptions of animal guile as an excuse for not catching them; we do not like looking incompetent and folklore loves a worthy opponent.

Secondly, for several decades there has been an accidental experiment playing out in Tasmania. There is almost unlimited food for thylacines. Numbers of wallabies have been steadily rising across developed areas, partly due to a dearth of predators but more in response to increased primary production. The abundance is spectacular on the edge of many farming districts. These are exactly the places thylacines were routinely found, because farms tended to be in productive areas and edges gave shelter with better hunting opportunities.

There are less Tasmanian devils than for many decades. Just as hyenas are a risk for young of big cats, devils would be a risk for thylacines pups. Early descriptions of where thylacine litters were found would have been easily accessible to devils. While thylacine and devils were at original numbers (say 2000 and 20 000 respectively), devil predation would be incidental and of no great population significance. Change that situation to where a few thylacines were just hanging on when devils were at their peak of ~60 000 and that incidental predation would become overwhelming. In much of the best thylacine habitat devils were at extraordinary densities for nearly 20 years until devil facial tumour disease (DFTD). Observing devils' omnipresence and persistence through their halcyon days I often wondered how on earth thylacines could raise pups.

Nobody is killing thylacines now. Snaring has ceased and the ubiquitous strychnine has long been replaced by 1080, greatly reducing the risks of secondary poisoning for native carnivores. Raptors, devils and quolls are very resistant to 1080 (essentially a plant defence) and it can reasonably be expected that the thylacine would also have natural resistance. There are furtive, whispered anecdotes of both accidental and deliberate killings of thylacines since the last known shooting by Wilf Batty in 1930 but there is no proof – none.

There have not been wild dogs in Tasmania for many decades. They became quite common in many places following their introduction by colonialists and were undoubtedly a direct threat to thylacines. Dingoes, after all, were likely key to thylacines' extinction on mainland Australia.

There remain many areas of refuge for thylacines in both wild areas and 'local wildernesses' – quiet corners that exist just a kilometre or two from people, places fitting all the early descriptions of dens.

So, we have a situation where food supply is way up, persecution and natural predation is way down and denning opportunities are more than adequate. One would expect a lag in recovery but how long? It's been five or more thylacine generations of optimal conditions, longer for adequate conditions. There's been plenty of time for dispersion and re-establishment in the old, best habitats; and for rediscovery.

Moments of media excitement, where people claim to have clear photos or film of thylacines, which do not prove to be so, are not evidence of thylacine whatever the enthusiasm of those concerned. The cyclic nature of this frenzy I think reflects a new craving for exciting good news in the face of woes about biodiversity loss, climate change and COVID-19.

It is fascinating to me that our dream of rediscovery and conservation is also being acted out on other sides of the world – in Japan and South America. The search for the Ezo wolf on the island of Hokkaido in central Japan perhaps has the closest parallels with our situation. That large, ancient wolf was for aeons closely enmeshed in local culture. There are hints of that starting here too, with our return of interest but so far these cultural aspects are unsophisticated and smack of Tasmanians' love of conspiracy theories. The Japanese wolf was also a very real animal that apparently went extinct ~140 years ago but sighting reports and inconclusive photos persist as do determined amateur searches using camera traps and field sign. I first encountered this effort nearly 40 years ago when having exchanges with Dr Luigi Boitani of the Wolf Working Group on detecting rare, large canids. One of the issues hotly debated was rewilding Japan with imported wolves to restore the ecological balance, perhaps a very rational reason to decide on extinction or not. We don't have that luxury.

Most exciting is research in the Amazon using cameras to find and study some of the world's most elusive and rare wild canids, the short-eared dog (aka the ghost dog) and the bush dog, oddly similar in appearance to the thylacine and Tasmanian devil respectively. To read the researchers emails and blogs one might be chatting to wildly enthusiastic thylacine searchers (surely a very high bar!). There is much in common with our efforts in dealing with eyewitness reports, but a key difference is of course that our South American friends have much confirmable, contemporary data.

Getting positive results helps maintain energy (and funding) and lets methods be improved. But, what is really gratifying is the professionalism and enthusiasm that leaps out of publications and websites. Indeed, one key paper has 42 authors, making it the longest author list by far I have ever seen (imagine accommodating everyone's comments on the draft!). That wide inclusion smacks of integrity. To me it's fantastic to see others living our dream; if not for COVID-19 I would visit just to enjoy how it might have been. Such a relief to see the human fascination with the fantastic being applied to the more ordinary.

That relief is also felt in Tasmania, with people finally paying proper attention to the Tasmanian devil, so distained when it was common. The real test will come as devils recover. Will we return to the bad old days of cavalier persecution or take a long overdue new path? Some stockowners on mainland Australia are even learning to appreciate dingoes, so maybe there is hope yet. I just regret not instituting a 'common species day' to rival our fascination for threatened species. Common species are, after all, arguably more what drives the ecology.

This situation playing out far away very much brings me back to the huge scholarship that is laid out in this book you are reading, some aimed at searches and status, others at reconstructing aspects of predatory ecology. The research in all its forms greatly contributes to our understanding of the thylacine and other such animals and the processes that imperil them.

Like everyone I have a temporary role in this saga. Sometimes overstatedly described as the subject's gatekeeper at TASPAWS (even being accused of hiding evidence to protect whatever interest the accusers railed against on the day), I was a filter for reports and carried out many a field assessment and reconstruction. One must remember that reporters may be right, they may have made a mistake, had an illusion or they may be lying. Modified memory is a real problem – our memories are not video replays. Pretty well every Tasmanian and many visitors know exactly what a thylacine looks like from advertising and branding, familiarity that can affect report details. Reconstructions of incidents were most informative. Reported distances and times were usually very optimistic. A distance reported as 100 m usually proved on measuring to be closer to 170 and 10 s became 2–3. Publications on thylacines never challenged reported details of sightings; in doing so, making some more credible than perhaps deserved.

In responding to hundreds of reported incidents over decades I confirmed all but the first option; even Hans Naarding's famous incident remains unconfirmed. But I can't deny that with some reports, including that one, I was left scratching my head, grinning. Maybe Loki does exist.

I know our South American friends are very mindful of the thylacine story. Let us hope this book helps them prevent a repeat.

Contents

Cultural sensitivity warning

Readers are warned that there may be words, descriptions and terms used in this book that are culturally sensitive, and which might not normally be used in certain public or community contexts. Although this information may not reflect current understanding, it is provided in a historical context.

This publication may also contain quotations, terms and annotations that reflect the historical attitude of the original author or that of the period in which the item was written, and may be considered inappropriate today.

Aboriginal and Torres Strait Islander peoples are advised that this publication may contain the names and images of people who have passed away.

Dedication

This book is dedicated to all those who have worked to understand or protect the thylacine over the years, whose efforts failed to prevent the inevitable, whose voices have been drowned out by those more raucous and less worthy and whose names have been all but forgotten or were never known beyond their long-suffering loved ones.

In the months following completion of this book, three respected members of the thylacine research community have sadly passed away: Ray Sawford, Col Bailey and Fillippa Buttita. Ray was a Hobart-based historian whose family had a connection with the thylacine that reached back to the 19th century. He carried with him a wealth of information on the interaction between the species and the people of the Midlands and successfully sought out images and references to the thylacine, generously sharing everything he discovered. Col Bailey, who passed away just weeks after the death of his beloved wife Lexia, needs no introduction. The author of numerous books and articles on the thylacine, he devoted decades of his life to searching for the 'Tasmanian tiger'. Arguably, his most valuable contribution was his interviews recording the final recollections of the last generation who definitely encountered the species. Filippa Buttitta was a talented and prolific artist who incorporated the thylacine into much of her work. Juxtaposing naturalistic images of the species into identifiably modern settings, Filippa's paintings are unmistakable and will always be admired among researchers and art lovers alike.

Branden Holmes and Gareth Linnard

Acknowledgements

We would first and foremost like to thank Briana Melideo (our Publisher) and Mark Hamilton (our Development Editor) at CSIRO Publishing for both their professional guidance and their personal patience. This volume took far longer than expected and they have wonderfully accommodated our project and supported it at every single step. We also thank Kerry Brown (our copy editor), who picked up many elements that had slipped past us and who did not hesitate to question if anything was unclear to her in the manuscript at certain points. This book is undoubtedly much better because of that.

Enormous thanks must go to Dr Stephen Sleightholme, who peer-reviewed all of the editors' own chapters to ensure that they were up to the same standards that we asked of our contributors. His comments on each draft were insightful and helped raise the quality and accuracy of our writing.

We are indebted to Nicole Dyble for her valiant attempts to find Indigenous contributors for the volume and her fantastic original artwork. Ann Lloyd-Jones at the Save the Tasmanian Devil Appeal also took it upon herself to try and find Indigenous contributors for the volume.

We are very grateful to the following people who contributed more than one chapter to this volume (number of chapters in brackets): Dr Stephen Sleightholme (6), Cameron Campbell (3), Nicole Dyble (2), Dr Nic Haygarth (2), Col Bailey (2), Dr Lauren White (2) and Dr Douglass Rovinsky (2).

We wish to thank the following people and organisations for permission to reproduce copyrighted material at no cost: Dr Stephen Sleightholme, Delphine Enkler, Anthony Black (Libraries Tasmania), the Tasmanian Police Museum, Dr Mikael Siversson (Western Australian Museum), Dr Ian Beveridge, Dr Peter White and Dr Peter Sheppard and the Museum of Victoria.

Finally, we wish to give a general and extended thanks to everyone who was involved with this project.

Branden Holmes and Gareth Linnard

Preface

> A brief review of the present position of the Tasmanian marsupials may be useful not only for its present interest, but as a source of reference to workers in future years who may endeavour to trace the extent and distribution of our fauna, many forms of which will undoubtedly become rare if not extinct. – Clive Lord, Existing Tasmanian marsupials (1927), p. 17.

Among other carnivores, the thylacine stands apart as unique. Aside from being the world's largest carnivorous marsupial, when it disappeared in the mid-20th century its taxonomic isolation ensured that it left no parallels through which we can easily comprehend it and it instead bequeathed us a series of contradictions.

It is a species with a high but multifaceted cultural profile, to many a symbol of extinction, to others one of hope. A unique emblem of its home state, but also a figurehead for all of Australia's wildlife. It has a historical record, both as a perceived scourge and as a high-value commodity. It lived recently enough to be the subject of an extensive and ever-expanding photographic record, but not a single colour image has ever been recovered and we have no recordings of its vocalisations. Our understanding of its ecology and behaviour is largely based on anecdote, yet its physical remains lend themselves to increasingly advanced and objective methods of study.

Propaganda too has definitely played a part in obscuring the thylacine, to an extent that led to the introduction of bounty schemes in the 19th and early 20th centuries, and, more insidiously, in historical misrepresentation, if not negationism, which concurrently proliferated with the expansion of the internet. Despite the advances made in the previous two decades it is sobering to consider that most of the information is presented to the public by non-specialist sources or else in user-generated internet content and needs to be addressed.

It is often said that the opportunity to study the species in the wild was squandered, but this is no truer of the thylacine than any other comparably rare species encountered during the 19th and early 20th centuries. The past was not peopled by the wilfully ignorant, but it could not offer the same opportunities as the present. Is there truthfully any species that has not become far better understood in the past few decades? The thylacine, a rare and crepuscular species, was neither ignored nor overlooked, but was rarely encountered and had disappeared before observational field biology became a widespread science.

Had modern scientific techniques, and equipment, been available to them, anyone familiar with the naturalists of early 20th century Australia could suggest a number who would most certainly, and avidly, have applied them to the study of the thylacine. But they did not have the opportunity and so our understanding of the thylacine is consequently deficient. No species is especially well described anecdotally, then or now.

The thylacine was undoubtedly persecuted, but name a carnivore that was not at that time; in fact, try to name one now, including those still being so treated in the affluent countries of the West. If a level of hostility toward the species ever truly existed to the degree it is popularly imagined today, it was short-lived and far from universal. It should be remembered the government bounty scheme, which actually only ran on crown land between 1888 and 1908, was passed in Parliament only by a single vote. The incentive it created to hunt thylacines was negligible, as demonstrated by Nic Haygarth's (2017) more nuanced picture of the incidental and opportunistic killing of thylacines.

We may also be overestimating the numbers of thylacines accounted for in the 1888–1908 scheme; on 27 September 1888 the Tasmanian News reported that its initial budget of £500 per annum was to be cut by £350, as so far only £10 had been paid (*Tasmanian News* 1888). At £1 per bounty, this accounts for 10 payments, yet Guiler (unpublished) records that 27 adult bounties had been paid by this time.

The mass digitisation of material, such as the National Library of Australia's free research portal, The Trove, launched in 2008, has served to greatly expand our knowledge of the historical record and to check many longstanding errors. More recently, two major projects have been undertaken to document all known photographs and films of the species and work is ongoing to tease out the scientific and historical details they contain. Eight new photographs have been discovered since the publication of the summary of the Thylacine Image Registry (Sleightholme and Campbell 2021a), and four new film sequences have been discovered since 2020, including the only known footage of a juvenile thylacine.

The thylacine itself has changed or perhaps more accurately it is being retrospectively returned to its true form. The species' average body mass has been significantly revised downwards from an anecdotal mean of ~30 kg, to an evidence-based 13.7 kg for females and 19.7 kg for males (Rovinsky *et al.* 2020), placing the thylacine below the threshold of 21 kg and under the upper size limit of a small prey specialist. Together with further post-2003 studies on the species' jaw and brain, this has provided much needed clarity regarding the thylacine's ecological niche. The thylacine's genome was published in 2018, which has fuelled research into the evolutionary relationships of the species (Feigin *et al.* 2018). Advancements in genetics are also now leading to questions of de-extinction; should we resurrect the species, could we and equally importantly, what would we get if we tried?

In view of the progress in this field, and some of the difficulties it faces, a book of this kind is rather overdue. It has been written to be of interest to those engaged, or personally immersed, in the thylacine, as well as offering an entry synthesis for those less familiar with the species.

No such thing as an expert

The field is now too wide for any one individual to be an expert. Rather than a single author overextending themselves by trying to authoritatively cover the entire subject, each section is authored by specialists in their field, all of whom have voluntarily provided their time and knowledge to raise funds and hopefully awareness of the world's current largest marsupial carnivore, the Tasmanian devil.

The number of researchers is not evenly distributed throughout the fields; consequently, some contributors appear more than once, particularly in the historical sections.

A further benefit of a multi-author volume is that it will inevitably contain contradictions. Each submission is the contributor's own and based on their interpretation of the available evidence. Where discussion overlaps between the authors this inevitably results in differences of interpretation. You will for example, read that the last captive specimen was trapped in 1930, 1931 and 1933, or most significantly that the thylacine itself is both extinct and extant. In each of these cases, only one position can be correct and in each instance one author is as convinced by their interpretation of the evidence as another. This is an honest reflection on the state of research; there is no consensus on many points and this is sometimes not reflected in non-specialist sources, which creates the false impression that a homogeneous view exists on the topics within the subject.

Omissions

The voices of Indigenous Australians are glaringly absent here and requires explanation. This omission is not the result of a lack of opportunity to be heard; each time members of the Aboriginal community was contacted the request was unsuccessful. While this is understandable, and is a decision respected by the editors and all involved in the production of this volume, it is one that is regretted in equal measure. This book is therefore acknowledged to be entirely from a Western perspective and should a second edition of this volume ever be attempted, the primary objective of a revision must be to substantially include the representation of Australia's First Nations people among the contributor list. The double extinction of Indigenous knowledge of a recently extinct species must and will be resisted.

Finally, the authors wish to acknowledge that within the past two decades, thylacine research has lost two of its most valued members, Dr Eric Guiler (1922–2008) and Professor Heinz Moeller (1936–2009). Both were instrumental in laying the foundations of what came after them. It may come as a surprise to some that much of the most significant research into the species has never been published in English, but is widely available in German.

Branden Holmes and Gareth Linnard
10 December 2021

Prologue

Greg Woods

'To lose one parent, Mr. Worthing, may be regarded as a misfortune; to lose both looks like carelessness.' This Oscar Wilde quote by Lady Bracknell to Jack (Ernest) Worthing was an ironic and sarcastic question directed at Mr Worthing, whose intent was to marry the daughter of Lady Bracknell. The tragedy was that, through no fault of his own, Jack Worthing was orphaned at an early age as he 'lost both parents'. Variations of this quote have been used to underscore other unfortunate situations, with an appropriate example being a paper by Hamish MacCallum and Menna Jones on Tasmanian devil facial tumour disease (McCallum and Jones 2006). A relevant modification would be, 'To lose one large marsupial carnivore may be regarded as a misfortune; to lose both would look like carelessness.' This marsupial carnivore adapted quote has a simple message: the Tasmanian devil *must* be protected.

Approximately 14 000 years ago Tasmania became isolated from mainland Australia. This provided a 'fortunate' opportunity that allowed the thylacine and the Tasmanian devil to be protected from the 'misfortune' that befell mainland thylacines and Tasmanian devils. The 'misfortune' was the 'loss' of both the thylacine and the Tasmanian devil approximately 3000 years ago. The 'loss' was potentially due to competition by dingoes, human impact and climate change (White *et al.* 2018b). In Tasmania, the 'misfortune' of extinction was again thrust upon the thylacine, this time forever. Human impact was the most likely cause of the thylacine's global extinction, but was it 'misfortune' or 'carelessness'? A common misconception was that thylacines were responsible for the mass killing of sheep, which was a terrible 'misfortune' as it led to a bounty on the thylacine (and Tasmanian devil). And contributed to the thylacine's ultimate demise and a drastic reduction in the Tasmanian devil population. Fortunately for the devil, but too late for the thylacine, the government introduced legislation to protect both marsupial carnivores.

The protection afforded to Tasmanian devils and the absence of competition from the thylacine led to a rapid population expansion. Low genetic diversity is a feature of the Tasmanian devil population, as it was for the thylacine (Menzies *et al.* 2012) and most likely arose from a genetic bottleneck caused by a small population that was able to expand (Morris *et al.* 2013). This reduced genetic diversity may not have had any severe consequences until 1996 and the unfortunate arrival of devil facial tumour disease (DFTD), first noticed by Christo Baars, a wildlife photographer. This 'misfortune' could have led to 'carelessness' had it not been for the dogged persistence of Menna Jones and Nick Mooney, who indefatiguably lobbied the government. It was not until 2003 that Australian and Tasmanian Governments, zoo organisations and the academic community established the Save the Tasmanian Devil Program. Lessons learned from the extinction of the thylacine prevented inaction, thereby avoiding 'carelessness' of the Tasmanian devil population.

An extraordinary collaboration occurred between various stakeholders with scientific and non-scientific backgrounds, leading to significant advances in understanding the remarkable biology of the Tasmanian devil and DFTD. These advances are captured in the award-winning book *Saving the Tasmanian Devil* (Hogg *et al.* 2019). The earliest advance in research into DFTD was that a virus did not cause the cancer, but that DFTD was a transmissible cancer (Pearse and Swift 2006). The cell of origin was a Schwann cell that first arose in a female devil and the cancer cells were transmitted as a consequence of the biting behaviour of devils. The consequence of the devil's low genetic diversity contributed to their susceptibility to DFTD (Patchett *et al.* 2020). However, the ability of the DFTD cancer cells to avoid the devil's immune system was most likely due to a combination of low genetic diversity and that the cancer cells do not express major histocompatibility complex molecules and are therefore virtually invisible to the devil's immune system (Siddle *et al.* 2007, 2013). DFTD was causing a dramatic decline in devil numbers and extinction was a realistic possibility (McCallum *et al.* 2009).

The establishment of insurance populations of Tasmanian devils at various locations around Australia ensured disease-free devil populations. However, the main objective is to ensure an enduring population of healthy wild Tasmanian devils, hence the need for scientific research. One way to ensure an enduring population and avoid 'carelessness' would be to make devils resistant to DFTD. An understanding of the devil's immune system would explain how the devil and its DFTD cancer interacted (Woods *et al.* 2015, 2018). As with human public health, vaccination would provide a pathway to implementing resistance.

However, research into a vaccine for DFTD has had mixed success. Initially, it was believed that devils would not generate an immune response to DFTD as no wild devils had recovered from DFTD and there was no evidence for any immune response. Diligent laboratory-based research accompanied by dedicated field studies provided promising evidence that a vast majority of devils immunised with modified DFTD cancer cells could produce an immune response against the cancer cells (Pye *et al.* 2018). However, protection from DFTD was, at best, temporary and more research is required to improve the vaccine and develop a cost-effective process whereby wild Tasmanian devils can be immunised. An oral bait vaccine has successfully reduced rabies' incidence and an oral bait vaccine is being researched as an approach to protect wild devils against DFTD (Flies *et al.* 2020).

Genetic sequencing advances have provided essential insights into the thylacine and the Tasmanian devil. For the extinct thylacine the insights were retrospective (Feigin *et al.* 2018), but for the Tasmanian devil the insights provided valuable research directions for preserving the devil and thus avoiding 'carelessness'. Molecular tools that arose from genomic sequencing led to the understanding that the DFTD cancer is remarkably stable and uncovered some of the genes that contribute to the proliferation of DFTD cancer cells (Kwon *et al.* 2020). Encouragingly, genetic studies provide hope for the Tasmanian devil in the form of an apparent molecular evolution in response to DFTD (Epstein *et al.* 2016). But without a doubt, one of the most encouraging signs has been the observation that some devils can recover from DFTD (Pye *et al.* 2016a). Potentially the devils themselves will 'evolve' to avoid 'carelessness'.

Despite the impressive advances in management and scientific research, the Tasmanian devil is still at risk. There are only two mammalian species affected by transmissible cancers (dogs and devils). Despite this rarity, the Tasmanian devil has the unique distinction of being affected by two different transmissible cancers. The second transmissible cancer, DFT2 appeared in 2014 (Pye *et al.* 2016b) and is also of Schwann cell origin (Patchett *et al.* 2019). Fortunately, DFT2 is restricted to a south-eastern corner of Tasmania, but clearly has the potential to spread, similar to DFT1 (the original DFTD) and increase the devil's extinction vulnerability. To allow this to happen would be 'careless' and concerted efforts must be implemented to prevent DFT2 from contributing to the demise of wild Tasmanian devils.

Research is an essential tool that will aid in the protection of Tasmanian devils, the world's largest carnivorous marsupial, a title inherited after the extinction of the thylacine. Donations made to the Save the Tasmanian Devil Appeal (www.utas.edu.au/devil) have been central to accruing vital funds that have been directed towards ground-breaking research. We are incredibly grateful for this support, as most of the references cited in this prologue would have benefited directly or indirectly from these funds. Furthermore, all royalties from the present volume will be donated to the Appeal, helping to ensure that research funds will be available to future researchers too.

Introduction: The thylacine in Australian ecosystems

Menna E. Jones

The ecological story of the thylacine (*Thylacinus cynocephalus*) epitomises the anguish of timing in the history of European colonisation and people's relationship with nature. The world's largest marsupial carnivore in historic times, the thylacine became officially extinct[i] in its last stronghold in Tasmania in about the 1930s (Guiler 1985). Another 50 years, with improving attitudes towards conservation, might have seen this tragedy averted. With its extinction, we lost an entire lineage of unique animals from Earth – the thylacine was the sole remaining member of the marsupial family Thylacinidae. Known by many different names to Aboriginal peoples in different parts of Australia, kaparunina is the name used in the composite language of palawa kani in Tasmania.[ii] Its scientific name means dog-headed (*cynocephalus*) pouched (*Thylacinus*, from the Greek thulakos) animal.

The thylacine was a dog-like carnivore that arose within the marsupials, mammals with pouches that came to dominate in the Southern Hemisphere. When animals with different origins evolve in isolation on different land masses, they evolutionarily converge in morphology to fulfil the range of ecological functions. The Australasian marsupial carnivores produced counterparts to the placental carnivores that came to dominate elsewhere, with a dog-like carnivore (the thylacine), a bone-crunching hyaena-like carnivore (the Tasmanian devil, *Sarcophilus harrisii*) and smaller flesh-eating and insect-eating carnivores (the quolls, *Dasyurus* spp.) (Jones 2003).

The thylacine was a rather slow-running animal that probably hunted by ambush and short pursuit (Jones 2003; Figueirido and Janis 2011), more the size and hunting style of a coyote (*Canis latrans*) than a wolf (*Canis lupus*), and more solitary than the similar-sized but pack-hunting dingo (Wroe *et al.* 2007). Thylacines have a long, narrow snout, more fox- than wolf-like (Jones and Stoddart 1998), although studies of skull biomechanics indicate that it produced high bite forces (Wroe *et al.* 2005). Thylacines are thought to have killed small to medium-sized prey (Jones and Stoddart 1998; Wroe *et al.* 2007; Attard *et al.* 2011), although European settlers in Tasmania reported them hunting kangaroos (Guiler 1985; Paddle 2000). The thylacine's canine or killing teeth are built to grab prey, not slash like wolves or dislocate with precision like large cats. They probably killed like devils and quolls by biting the nape or chest and hanging on until the prey succumbed (Jones and Stoddart 1998). Thylacines lack the robust bone-cracking molar teeth of devils and probably did not eat the tough parts of carcasses such as bones (Jones 2003).

A litany of extinctions led to the demise of the final thylacinid carnivore. The family Thylacinidae arose in the Oligocene (33.9–23 Mya) and by the Miocene (23–5.33 Mya) there were many species, ranging from cat- to wolf-sized (Archer 1982; Long 2002). Thylacinids were gradually replaced by the dasyurid carnivore lineage (family Dasyuridae), of which there

---

i https://www.iucnredlist.org/species/21866/21949291
ii http://tacinc.com.au/three-capes-welcome/

are now 66 species (Woinarski *et al.* 2014), until there was just *T. cynocephalus* remaining in the Holocene (Archer 1974). The species originally roamed across Tasmania, Australia and New Guinea, the land masses that were joined together to form Sahul (or Greater Australia) during the Pleistocene (2.58 Mya–11.7 kya).

Thylacines and Tasmanian devils became extinct on the Australian mainland ~3000 years ago, probably from a synergy between a long period of warm, dry climate (El Niño) that reduced populations and genetic diversity (Brüniche-Olsen *et al.* 2018; White *et al.* 2018a), combined with competition with the newly introduced and larger dingo (Fillios *et al.* 2012; Letnic *et al.* 2012), and increasing human populations (Prowse *et al.* 2014). The thylacine survived the European invasion of Tasmania (1803) but declined precipitously a century later between 1906 and 1910 (Guiler 1985). The cause of their final population collapse is still debated; it followed decades of persecution as sheep killers (Guiler 1985), but a hypothesis of infectious disease, potentially disease spillover from domestic dogs, cannot be ruled out (McCallum and Dobson 1995), given the sudden decline simultaneously across Tasmania and anecdotal records of simultaneous deaths from a 'distemper-like' disease (Guiler 1961).

Approaching 100 years now since the last thylacine was captured in the wild, attitudes have changed. Without persecution, thylacines would do well in Tasmania. Forest clearing and fragmentation for agriculture benefit their macropod prey (Driessen and Hocking 1992), supporting an increased abundance of devils in the late 1900s (Patton *et al.* 2019). We don't know how the loss of the thylacine affected devils or quolls because we cannot disentangle in time thylacine extinction from landscape change. Larger predators persecute smaller predators, often killing them, and there is evidence of the competitive influence of the thylacine on the smaller marsupial carnivores.

With the mainland extinction of the thylacine and devil, rapid evolution occurred in the tooth morphology of the spotted-tailed quoll and the eastern quoll, as they relaxed from competition to eat larger prey with their now larger canine teeth (Jones 1997). The thylacine was the top mammalian predator in Australian ecosystems prior to the introduction of the dingo. Although there are many differences between thylacines and canids, their ecological role may be sufficiently similar that the dingo has functionally replaced the role of the thylacine on mainland Australia (see Wroe *et al.* 2007), with no known changes in macropod prey that can be attributed to the loss of thylacines.

Sometime between 1977 and 1987 (Patton *et al.* 2020), a female devil in north-east Tasmania developed a tumour on the face that developed from a Schwann cell in the myelin sheath that surrounds nerves (Murchison *et al.* 2010). Devils' faces are highly sensory, dense with nerve cells and a rich array of thick black vibrissae or whiskers, all the better to 'see' in the dark. Devils sustain many injuries on their face and inside their mouth from biting each other (Hamede *et al.* 2012). Perhaps the extremely high population density of devils in the 1980s (Patton *et al.* 2019) increased the chance of this tumour evolving into a rare transmissible cancer, named devil facial tumour disease (DFTD), in which a cancer cell takes the evolutionary step of metastasising outside the host body to infect other individuals (Pearse and Swift 2006). In the same year that DFTD was diagnosed, my colleague Hamish McCallum and I published a decision tree for how to proceed with managing and researching emerging infectious diseases in wildlife that are poorly known (McCallum and Jones 2006), the paper title borrowing the words Oscar Wilde put in Lady Bracknell's mouth 'To lose one marsupial carnivore may be regarded as a misfortune; to lose both would look like carelessness'.

The rest of this history of the devil and its tumour disease will end on a happier note than that of the thylacine. Despite early concerns that DFTD could cause extinction (McCallum *et al.* 2009), no local populations have become extinct even though the disease has now spread to most of the devil's range (Cunningham *et al.* 2021). Devils are rapidly evolving resistance and tolerance to the disease (Epstein *et al.* 2016; Ruiz-Aravena *et al.* 2018) and although broad-scale population recovery is not yet evident, individuals are recovering (Pye *et al.* 2016a), some to live to old age and others living for years with slow-growing tumours (Jones *et al.* 2019). Recent modelling indicates that the disease has entered an endemic state and that population decline, still driven by geographic spread, will stabilise within 5 years (Patton *et al.* 2020; Cunningham *et al.* 2021).

The cascading effects of the loss of the devil from facial tumour disease indicates the extent of the changes that might have occurred when the thylacine was lost. Using two large natural experiments, the progressive east to west decline of devil populations and introduction of the devil as a conservation measure to Maria Island off the east coast of Tasmania, we can measure cascading changes in the mammalian community (Cunningham *et al.* 2018, 2019a, 2019b, 2020). Devils appear able to suppress populations of feral cats and black rats, both highly destructive invasive species (Cunningham *et al.* 2020; Scoleri 2020). Without functional populations of devils, high densities of cats can have a devastating effect on wildlife (Hamer *et al.* 2021) and Tasmania is losing its small and medium-sized mammals, with animals such as eastern quolls and bandicoots increasingly in trouble (Fancourt *et al.* 2015; Cunningham *et al.* 2020).

What is the prognosis for the restoration of Tasmania's and Australia's mammalian communities and ecosystems? In the absence of the dingo and foxes, Tasmania has among the most intact mammal community in Australia. Tasmania also has the most diverse community of marsupial carnivores in the world, still with three species: the devil and two species of quolls. Extinction is now considered to be the least likely outcome of the facial tumour epidemic (Wells *et al.* 2019). Recovery of the devil seems likely; over what time-frame is uncertain, although rapid evolution of resistance has occurred in as little as 5 years following disease outbreak in local populations (Epstein *et al.* 2016) and the disease turned from epidemic to endemic in 25 years from its period of rapid spread in the mid-1990s (Patton *et al.* 2020).

Devil recovery should restore top-down control of cats and rats and protect smaller native species. The current focus of our conservation and restoration management and research is to restore structurally complex habitats and test ways to reduce cats by controlling their hyper-abundant European rabbit prey to enable native wildlife to live in the landscape with feral cats. Although analyses indicate that the rediscovery of the thylacine is extremely unlikely, given its size and the amount of time since it was last confirmed to be alive (Fisher 2010), we do not know exactly when a species goes extinct. I hope the thylacine has managed to survive somewhere in Tasmania, maybe in a harsh and remote area where a few always lived. We bear the legacy of societal attitudes towards wildlife, and for a few decades of history. It would be incredible to see the remarkable thylacine back in its place as the world's top marsupial carnivore.

The 80 short chapters in this volume attest to the rapidly growing body of research into the thylacine, including its evolution, anatomy and ecology, Indigenous knowledge of the species, how and why it was lost, and the possibility that it survived beyond 1936. And they emphasise the importance of an interdisciplinary approach to understanding ecology, extinction and their relationship. Species do not exist in a vacuum and changes in the fortunes of one species invariably means changes to the ecosystem, especially when that change is the extinction of a top predator such as the thylacine.

Interest in the species is perennial and is here being used to help raise funds for the protection and study of its smaller cousins that have thus far escaped the same fate, though the threat of complacency is ever-present, as the iconic loss of the thylacine reinforces. Although itself extinct, the thylacine has much to teach us about living species.

List of contributors

Daisy Ahlstone
Department of Comparative Studies, Ohio State University, Columbus, OH 43210, USA

Ken W. S. Ashwell
Department of Anatomy, School of Medical Sciences, University of New South Wales, Sydney, NSW 2052, Australia

Christopher Atkinson
Department of Gastroenterology, University of New Mexico School of Medicine, Albuquerque, NM 87106, USA

Marie R. G. Attard
Department of Biological Sciences, School of Life and Environmental Sciences, Royal Holloway University of London, Egham, Surrey TW20 0EX, UK; School of Engineering and Innovation, Open University, Milton Keynes MK7 6AA, UK

Jeremy J. Austin
The Environment Institute and School of Biological Sciences, University of Adelaide, Adelaide, SA 5005, Australia

Col Bailey
Tasmanian Tiger Research and Data Centre, New Norfolk, TAS 7140, Australia

Guy-Anthony Ballard
School of Environmental and Rural Science, The University of New England, Armidale, NSW 2351, Australia; University of Technology Sydney, Faculty of Science, Ultimo, NSW 2007, Australia; Centre for Inflammation, Centenary Institute, Sydney, NSW 2050, Australia

Peter B. Banks
School of Life and Environmental Sciences, University of Sydney, Sydney, NSW 2052, Australia

Robin Beck
School of Science, Engineering & Environment, University of Salford, Salford M5 4WT, UK

Katherine Belov
School of Life and Environmental Sciences, The University of Sydney, Sydney, NSW 2006, Australia

Gregory S. Berns
Psychology Department, Emory University, Atlanta, GA 30322, USA

Anthony Black
Tasmanian Archive and Heritage Office, Hobart, TAS 7000, Australia

Corey J. A. Bradshaw
College of Science and Engineering, Flinders University, Adelaide, SA 5001, Australia

Kaye Brown
Department of Anthropology, Boston University, Boston, MA 02215, USA

Andrew A. Burbidge
Floreat, WA 6014, Australia

Aaron Camens
College of Science and Engineering, Flinders University, Bedford Park, Adelaide, SA 5042, Australia

Cameron R. Campbell
Eagle Mountain Circle, Fort Worth, TX 76135, USA

Erica A. Cartmill
Department of Anthropology, University of California, Los Angeles, CA 94720, USA; Department of Psychology, University of California, Los Angeles, CA 90095, USA

Matt Cartmill
Department of Anthropology, Boston University, Boston, MA 02215, USA; Department of Evolutionary Anthropology, Duke University, Durham, NC 27710, USA

John Doyle
Weatherstone Circuit, Googong, NSW 2620, Australia

Nicole Dyble
Tasmanian Devil Experience, Saffire Freycinet Luxury Lodge, Coles Bay, TAS 7215, Australia

Russell K. Engelman
Department of Biology, Case Western Reserve University, Cleveland, OH 44106, USA

Charles Y. Feigin
School of BioSciences, The University of Melbourne, Parkville, VIC 3052, Australia; Department of Molecular Biology, Princeton University, Princeton, NJ 08544, USA

Borja Figueirido
Departamento de Ecología y Geología, Facultad de Ciencias, Universidad de Málaga, Málaga, ES-AN 29071, Spain

Peter John Sabine Fleming
New South Wales Department of Primary Industries, Orange, NSW 2800, Australia

Richard Freeman
Centre for Fortean Zoology, Woolfardisworthy, Bideford EX39 5QR, UK

Katie Glaskin
Goolugatup Heathcote Cultural Precinct, Applecross, WA 6153, Australia

Daniel Gonzalez-Socoloske
Department of Biology, Andrews University, Berrien Springs, MI 49104-0410, USA

Tammy Gordon
Independent Researcher, Launceston, TAS 7250, Australia

Catherine Grueber
School of Life and Environmental Sciences, Faculty of Science, The University of Sydney, Camperdown, NSW 2006, Australia

Tony Harper
DeBusk College of Osteopathic Medicine, Lincoln Memorial University, Knoxville, TN 37932, USA

Adam Hartstone-Rose
Department of Biological Sciences, North Carolina State University, Raleigh, NC 27695, USA

Nic Haygarth
Heritage Tasmania, GPO Box 618, Hobart TAS 7000, Australia

Mieke van der Heyde
Trace and Environmental DNA Laboratory, School of Life and Molecular Sciences, Curtin University, Perth, WA, 6102, Australia; ARC Centre for Mine Site Restoration, School of Molecular and Life Sciences, Curtin University, Bentley, Perth, WA, 6102, Australia

Dieter F. Hochuli
School of Life and Environmental Sciences, The University of Sydney, Sydney, NSW 2006, Australia

Carolyn Hogg
School of Life and Environmental Sciences, The University of Sydney, Sydney, NSW 2006, Australia

Branden Holmes
PO Box 21, Two Rocks, WA 6037, Australia

Christine M. Janis
Bristol Palaeobiology Group, School of Earth Sciences, University of Bristol, Bristol, BS8 1RJ, UK; Department of Ecology and Evolutionary Biology, Brown University, Providence, RI, 02912, USA

Menna E. Jones
School of Natural Sciences, University of Tasmania, Hobart, TAS 7001, Australia

Shimona Kealy
School of Culture, History and Language, College of Asia and the Pacific, Australian National University, Canberra, ACT 2601, Australia; ARC Centre of Excellence for Australian Biodiversity and Heritage, Australian National University, Canberra, ACT 2601, Australia;

Evolution of Cultural Diversity Initiative, Australian National University, Canberra, ACT 2601, Australia

Jamie B. Kirkpatrick
School of Geography, Planning and Spatial Sciences, University of Tasmania, Hobart, TAS 7001, Australia

Tessa Knights
Australian Research Centre for Human Evolution, Griffith University, Nathan, Brisbane, QLD 4111, Australia

Mackenzie Kwak
Department of Biological Science, National University of Singapore, Singapore 117558

Rebecca Lang
Strange Nation Publishing, Wentworth Falls, Sydney, NSW 2782, Australia

Michelle C. Langley
Australian Research Centre for Human Evolution and Archaeology, School of Environment and Science, Griffith University, Nathan, Brisbane, QLD 4111, Australia

Chris Lee
Priestley College, Warrington, Cheshire, WA4 6RD, UK

Gareth Linnard
Oak Drive, Woodland Park, Waunarlwydd, Swansea, SA54QP, UK

Kathryn Medlock
Tasmanian Museum and Art Gallery, Hobart, TAS 7000, Australia

Paul Meek
NSW Department of Primary Industries, PO Box 530, Coffs Harbour, NSW, Australia; School of Environmental and Rural Science, University of New England, Armidale, NSW, Australia

Brandon R. Menzies
School of BioSciences, The University of Melbourne, Parkville, VIC 3010, Australia

Carly Monks
Department of Archaeology, The University of Western Australia, Crawley, Perth, WA 6009, Australia

Nick Mooney
Tasmanian Museum and Art Gallery, Hobart, TAS 7000, Australia

Mary-Jane Mountain
School of Archaeology and Anthropology, Australian National University, Canberra, ACT 2601, Australia

Ken Mulvaney
Centre for Rock Art Research and Management, The University of Western Australia, Perth, WA 6009, Australia; Heritage Team, Rio Tinto, Dampier, WA 6713, Australia

Axel Newton
School of BioSciences, The University of Melbourne, Parkville, VIC 3052, Australia

Winston Nickols
Penguin History Group, Penguin Railway Station, Penguin, TAS 7316, Australia

Julie M. Old
School of Science, Hawkesbury Campus, Western Sydney University, Penrith, NSW 2751, Australia

Andrew Pask
School of Biosciences, University of Melbourne, Parkville VIC 3052, Australia; Department of Sciences, Museums Victoria, Carlton, VIC 3053, Australia

Emma Peel
School of Life and Environmental Sciences, The University of Sydney, Sydney, NSW 2006, Australia

Cassia Piper
Department of Earth and Planetary Sciences, Western Australian Museum, Welshpool, WA 6101, Australia; Biodiversity Information Office, Department of Biodiversity, Conservation and Attractions, Kensington, WA 6151, Australia

Douglass S. Rovinsky
Department of Anatomy and Developmental Biology, Biomedicine Discovery Institute, Monash University, Clayton, VIC 3800, Australia; College of Science and Engineering, Flinders University, Bedford Park, SA 5042, Australia

Frédérik Saltré
Global Ecology Lab, College of Science and Engineering and ARC Centre of Excellence for Australian Biodiversity and Heritage, Flinders University, Adelaide, SA 5001, Australia

Mikael Siversson
Collections and Research Centre, Western Australian Museum, Welshpool, WA 6106, Australia

Stephen R. Sleightholme
Bitham Mill, Westbury, BA13 3DJ, UK

Steffen Springer
SRH Wald-Klinikum Gera, Straße des Friedens 122, D-07548 Gera, Germany

Chris Tangey
Alice Springs Film and Television, Alice Springs, NT 0871, Australia

Kailah M. Thorn
Edward de Courcy Clarke Earth Science Museum, School of Earth Sciences, University of Western Australia, Crawley, Perth, WA 6009, Australia; Department of Earth and Planetary Sciences, Western Australian Museum, Welshpool, WA 6106, Australia

Kenny J. Travouillon
Collections and Research Centre, Western Australian Museum, Welshpool, WA 6106, Australia

Karl Vernes
Ecosystem Management, University of New England, Armidale, NSW 2351, Australia

Peter Veth
Centre for Rock Art Research and Management, The University of Western Australia, Perth, WA 6009, Australia; Australian Research Council Centre of Excellence for Australian Biodiversity and Heritage, University of Wollongong, Wollongong, NSW 2500, Australia

Michelle Vickers
Today's Psychology, Brendale, Brisbane, QLD 4500, Australia

Liana F. Wait
Department of Ecology and Evolutionary Biology, Princeton University, Princeton, NJ 08544, USA

Natalie M. Warburton
Harry Butler Institute, Murdoch University, Murdoch, WA 6076, Australia; Department of Earth and Planetary Sciences, Western Australian Museum, Welshpool, WA 6106, Australia

Lauren C. White
Australian Centre for Ancient DNA, School of Biological Sciences, University of Adelaide, Adelaide, SA 5005, Australia; Arthur Rylah Institute for Environmental Research, Department of Environment, Land, Water and Planning, Heidelberg, VIC 3084, Australia

Mike Williams
Strange Nation Publishing, Wentworth Falls, Sydney, NSW 2782, Australia

Greg Woods
Menzies Institute for Medical Research, College of Health and Medicine, University of Tasmania, Hobart, TAS 7000, Australia

Stephen Wroe
School of Environmental and Rural Science, University of New England, Earth Sciences Building, Armidale, NSW 2351, Australia

Michael Zieger
SRH Wald-Klinikum Gera, Straße des Friedens 122, D-07548 Gera, Germany

PART 1: ANATOMY, BIOLOGY AND ECOLOGY

The International Thylacine Specimen Database

Stephen R. Sleightholme

Preserved within the darkened store rooms of some of the world's major museum and university collections are all that physically remains of a species now presumed extinct – the thylacine or Tasmanian tiger (*Thylacinus cynocephalus*). First published in 2005, and now into its seventh revision, the International Thylacine Specimen Database (ITSD) contains the detailed records of 812 known specimens, comprising skulls, skeletons, skins, organs and the preserved bodies of adults and pups. These specimens are held in 118 collections in 23 countries, with a small number in private ownership.

Natural history museums traditionally collect, classify, conserve, study and exhibit biological specimens. The value in creating these collections is in the accompanying data, detailing when and where a specimen was collected, and by whom. Without this contextual information, a specimen's value to science is considerably diminished.

Taxonomy was the pioneering science of the 19th century and to a great extent it galvanised the race to acquire as many new specimens as possible. A diverse coterie of collectors and animal dealers were charged with procuring thylacines for the various institutions and in their frenzied pursuit to obtain specimens, valuable collection data were frequently omitted, lost or destroyed. Consequently, detail within the records for many of the specimens in the ITSD is sparse, to say the least, with most entries simply noting Tasmania as the geographical source.

A thriving commercial trade in thylacine specimens flourished from the middle of the 19th century and continued unabated until the late 1920s. The very nature and conduct of this enterprise undoubtedly contributed to the scarcity of source data within today's collections. Many of the major European sea ports had their assortment of specialist dealers in natural history and curios. Names such as Salmin, Frank, Jamrach, Umlauff, Frič, Reiche and Leadbeater appear frequently throughout the ITSD as the source of supply of a significant number of thylacine specimens. These dealers bought and traded shells, skins, skeletons, eggs, fossils, minerals and birds from the crews of returning vessels and sold them to amateur naturalists, museums and universities. Few of the specimens purchased from these dealers were accompanied by collection data and are simply recorded in museum registers with only a date and the named supplier.

Within Tasmania, significant numbers of thylacines were procured through local agents who purchased specimens (dead or alive) through classified advertisements in the local press. A thylacine was therefore traded like any other commodity, initially being sold to the agent by a land owner, farmer or trapper and then resold to museums and universities as specimens. Unsurprisingly, few of these commercially sourced specimens were accompanied by any data.

In the late 1800s, the Tasmanian museums began to appreciate that the rarity of the colony's native animals could be put to good advantage in building their fledgling collections through exchanges with other museums around the world. They procured an unknown number of thylacines (some by means of government bounty kills) for use in exchanges, but again, few of these specimens were accompanied by any source data.

In the continuing development of the ITSD, the task of retrospectively piecing together the missing provenance of the specimens has been challenging, but with a greater understanding of the background of thylacines that entered zoos and the histories behind many of the captures and kills, the detailed provenance of a good number of the specimens is now known.

Island homes such as Tasmania are nature's arks and often the last bastion for a species. The natural barrier of the sea protects them from introduced animal pests and diseases and, in some respects, from humankind itself. Specimens of the thylacine serve as poignant reminders of the fragility of such populations and the relative ease with which, through misguided judgements, they can be destroyed.

Sir Colin MacKenzie's remarkable legacy

Stephen R. Sleightholme

Preserved within the wet specimen collection of the National Museum of Australia is the remarkable legacy of Sir Colin MacKenzie (1877–1938). MacKenzie was a distinguished, Melbourne-based orthopaedic surgeon who devoted much of his life to the study of Australian fauna. In 1919, he formed and financed the Australian Institute of Anatomical Research and began building a collection of preserved specimens of Australian wildlife. This work intensified in the 1920s when the Victorian Government granted him permission to establish a field research station at Healesville that enabled him to breed and collect native animals for use as anatomical specimens. MacKenzie states: 'A knowledge of these animals is absolutely essential for a thorough knowledge of the human body. They are teeming with points of vital interest to ourselves and worthy of something better than to be slaughtered for the sake of sport or skins' (*The Daily Telegraph* 1924).

In 1923, MacKenzie generously donated his entire collection of marsupial specimens, including those of the thylacine, to the Australian nation. The following year, the Government responded by creating the National Museum of Australian Zoology to house them, appointing him as its first director. In 1931, the museum became known as the Australian Institute of Anatomy (AIA) to coincide with the opening of its Canberra home. The AIA closed in December 1985 and the collection transferred to the National Museum of Australia. MacKenzie's bequest is an exceptional resource, and comprises just under half of the total 70 entries for organ specimens within the sixth revision of the International Thylacine Specimen Database (ITSD).

MacKenzie was actively involved with the direction of the Melbourne Zoo and the majority of his thylacine specimens were obtained postmortem from animals that died at the zoo. Paddle (2012) notes that at least six of Melbourne's thylacines were his own purchases, loaned to the zoo to complete the remainder of their lives in captivity. In an article published in the Brisbane newspaper *The Telegraph*, reference is made to one of MacKenzie's thylacines and to its final fate as one of his specimens:

> Here is a marsupial Tasmanian wolf, now a very valuable animal, as it is rapidly becoming extinct. It is the property of Professor Colin MacKenzie, director of the National Museum of Australian Zoology, and will eventually make its home in Canberra (*The Telegraph* 1927).

MacKenzie left specific instructions for when an animal died. In the minutes of the Zoological and Acclimatisation Society of Victoria dated February 1922, it states: 'Chief secretary wrote that facilities should be given to Dr. MacKenzie that, when animals die, a phone message should be sent at once and a room set apart for post-mortem examinations, and a small tank containing some formalin be provided'.

In addition to the Melbourne Zoo specimens, MacKenzie is also known to have purchased the body of at least one thylacine from James Harrison, the wild animal dealer based in Wynyard, on the north-west coast of Tasmania. Harrison's notebook records that on 11 May 1928, he supplied a 'tiger' to MacKenzie for the sum of £12 (Harrison unpublished). The specimen was a juvenile that had died in Harrison's care. At the same time, MacKenzie purchased a live specimen from Harrison, which was forwarded to the Melbourne Zoo. During the 1920s, MacKenzie also purchased the bodies of at least three thylacines that died at the Beaumaris Zoo. In total, it is estimated that MacKenzie acquired at least 12 thylacines as source material for his organ preparations.

MacKenzie's work focused entirely on the anatomical attributes of his specimens. Unfortunately, this constraint reduced their value from a scientific perspective, as he failed to keep detailed records for each specimen, only labelling them with a brief description of the organ preserved and the common name of the animal donor. That said, MacKenzie's contribution to our knowledge of the internal anatomy of the thylacine was the foresight to collect and preserve all of its internal organs for future generations of scientists to study.

MacKenzie (1924) poignantly wrote of the thylacine and other rare native fauna: 'Unfortunately these animals are fast disappearing, and, in less than twenty years it is computed, will, in the absence of rigid protective measures, be all extinct'.

The fate of London Zoo's last thylacine

Stephen R. Sleightholme and Cameron R. Campbell

The Regent's Park Zoo in London was the first zoo in the world to exhibit thylacines. Outside of Australia, it exhibited more thylacines than any other zoo, with a total of 20 animals on display between 1850 and 1931, the longest time span of any zoo.

The last of London's thylacines was obtained from the Beaumaris Zoo (Hobart) in August 1925 by the animal dealer, Bruce Chapman, in exchange for an elephant and 30 Bennett's wallabies (Fig. 1). Two females made up the exchange, but due to protracted strike action by British seamen, the ship transporting them to London was forced to remain at sea for nearly 6 months. During this time, one of the females died. The *Daily Mercury* (1926) notes:

> One of the rarest of living animals - a Tasmanian wolf, or Thylacine - has arrived at the London Zoo. The animal is now so rigidly protected by the authorities, however, that the London dealer who imported the wolf was not allowed to take a male out of the country, and had to content himself with two females. Owing to the shipping strike, the two voyagers were six months aboard ship. One of them died, but the survivor is in fine condition, and, being the only specimen of its kind in a European Menagerie, and the last, it is said, to be allowed to be exported alive, it is not likely to suffer from lack of attention.

The Zoological Society purchased the surviving female from Chapman for £150 (Edwards unpublished) and she was placed on display in the North Mammal House from 26 January 1926, until her death on 9 August 1931. She became, and still remains, the last of her kind to be displayed outside of Australia.

Immediately following her death, she was injected with preservatives for anatomical study. The body was then dissected and the head sent to Professor Edwin Goodrich at Oxford University in September 1931 and the torso to Professor William Rowan at the Natural History Museum in Edmonton (Canada) the following month.

Bernard William Tucker (1901–1950), Demonstrator in Zoology and Comparative Anatomy at Oxford University, was charged by Goodrich with the detailed dissection of the

Fig. 1. Young female thylacine at London Zoo 1926, by F. W. Bond. Courtesy: Zoological Society of London.

head and neck. He was astonished, when asked to undertake the study, to find that virtually nothing had been previously written on the subject:

> Considerable attention has been paid to the head and neck musculature of marsupials and naturally a certain amount of information on the cranial nerves and blood vessels is scattered throughout the literature, but nowhere could I find the kind of coherent and comprehensive account of these portions of the anatomy of a marsupial type which I had expected would afford a basis of comparison with Thylacinus. No doubt a partial explanation of this astonishing state of affairs is that morphologists have found the cranial anatomy of the marsupials so closely similar to that of the placentals. This similarity, indeed, is well known. Yet it seems very strange and highly unsatisfactory that there should be any part of the anatomy of a whole sub-class of mammals of which a full and connected account is not available. This consideration, together with the fact that the extreme rarity of Thylacinus seemed to render it almost a duty to make the fullest possible use of the material, convinced me of the need to make a more detailed study than I had first visualised (Tucker unpublished).

Tucker proceeded to investigate the arterial, venous and nerve supplies and produced detailed notes and working drawings of his findings. In addition, he also wrote a comprehensive account of the musculature of the neck and shoulder.

In the 1970s, Professor Mike Archer of the University of New South Wales intended to resume the dissection where Tucker had left off. On receiving the head from Oxford, he found that virtually all of the muscles had been removed, rather than simply being resected. Consequently, he found that he was unable to validate any of Tucker's original research or countenance continuing with the dissection. However, he did make comment on one unusual discovery:

> What was bizarre, however, was the fact that in each orbit of the Thylacine head there was a dead mouse curled up where the Thylacine's missing eyeballs had been. Bernard Tucker must have had a weird sense of humour, or a mischievous assistant (M Archer, *pers. comm.*).

Tucker's unpublished notes and drawings, *A Contribution to the Anatomy of Thylacinus*, are held at the University Museum of Natural History in Oxford, together with the remains of the dissected head, the skin of which is the finest example in existence (Plate 1).

As far as is known, William Rowan did not record or publish his findings on the dissection of the torso (Rowan unpublished). However, he did exchange a headless skin, an incomplete skeleton and body parts in formalin, for other unnamed specimens with the Harvard Museum of Comparative Zoology (MCZ) in 1938. Although Rowan's thylacine specimens are catalogued together (36797), it now appears that the skull constituting part of the exchange was sourced from a different animal.

Attached to the rear leg of the skin (Plate 2) is an old identification tag with the numbers '612/31' written in ink, the '612' referring to Tucker's original specimen reference number and '31' to the year of receipt. The MCZ accession card also notes that: 'body parts in formalin were loaned to and returned by R. E. Rewell (Zoological Society of London) at Prof. Rowan's request,' and concludes by stating: 'these should not be discarded without consulting both of these people'. Dr Rewell was the senior pathologist at the London Zoo; however, his specific research interest in the thylacine 'body parts' is unknown. As of September 1982, these 'wet' specimens could no longer be located in the museum's collection and most likely were destroyed.

The year date (1931), the absence of the head on the skin, the connection to Rowan and the reference to the body parts being loaned to Dr Rewell at the Zoological Society in London, all firmly support the Harvard skin and the accompanying post-cranial skeleton belonging to the London Zoo's last thylacine. The lasting legacy of this thylacine will undoubtedly remain Tucker's unpublished study of the species' cranial anatomy, the most detailed ever produced.

Our growing knowledge of thylacine pouch young development

Julie M. Old

As the thylacine (*Thylacinus cynocephalus*) is extinct, we therefore know very little about its unique biology, including rate of growth and specific development and immune system. We can now only surmise, and presume, that growth and development and immune system function occurred in a similar manner to other marsupials. Further, the paucity of specimens now able to be examined adds to this difficult task of uncovering more about the species' immune system.

One pouch young specimen, from a litter of four, of unknown age, has been serially sectioned and is currently kept at the Museum of Victoria in Melbourne (Sleightholme *et al.* 2012). The histological sections were examined and a description of the histological appearance of the tissues recorded (Old *et al.* 2015). Based on the similarities in size and external physical features of other marsupial pouch young, particularly the closely related Tasmanian devil (*Sarcophilus harrisii*), as well as the histological appearance of the immune tissues, the age of the pouch young thylacines was extrapolated and the key developmental stage hypothesised.

The head length measurement of marsupial pouch young incrementally increases for the first 50 days of pouch life (e.g. Lyne and Verhagen 1957). As an adult Tasmanian devil weighs around half the body weight of an adult thylacine (van Dyck and Strahan 2008), it suggests a newborn thylacine would have had a head length measurement approximately twice that of a newborn Tasmanian devil pouch young. Tasmanian devil newborns have a head length measurement of 2.8 mm (L. Hughes, *pers. comm.*), hence thylacine newborns would have had a head length measurement of approximately 5.6 mm. Based on this combined information, a head length growth curve was able to be extrapolated for thylacine pouch young and, using this curve, the thylacine pouch young age was estimated to be approximately 39 days postpartum (i.e. since birth), based on the mean head length (36 mm) recorded for the three intact thylacine pouch young siblings in Sleightholme *et al.* (2012). Furthermore, based on the graph devised by Tyndale-Biscoe (2005) it suggests a newborn thylacine would have weighed approximately 600–650 mg.

An examination of the immune tissues, lungs and kidneys of the sectioned thylacine pouch young was conducted. The thymus, a primary immune tissue responsible for production of T cells and hence essential for immune defence, appeared mature and had not yet undergone any atrophy (i.e. reduction in size). We can therefore confirm that histologically, as with other marsupials (Old and Deane 2000; Borthwick *et al.* 2014), the thylacine's thymus is the first immune tissue to mature. Also, only a thoracic thymus was present, confirming the findings of Johnstone (1898), as observed in other carnivorous marsupials (Haynes 2001). Areas of blood cell development were visible in the bone marrow spaces and some showed adipocyte presence. The liver was mature in appearance and no areas of blood cell development remained, providing further evidence these were not newborn thylacines (Old 2016).

The spleen was not yet mature in histological appearance, with early follicles just starting to develop in the white pulp areas; hence it was not possible to confirm if the thylacine spleen acts as a major red blood cell storage site. No mucosal-associated lymphoid tissue was observed in the intestines or lungs. The lungs and kidneys were not yet mature in histological appearance. The immaturity of the lungs in particular was similar to that of other pouch young marsupials (e.g. Gemmell 1986; Runciman *et al.* 1996) and provides support for the theory that these thylacine young had not yet exited the pouch.

Examination of the gastrointestinal tract provided additional insights into the developmental stage of the thylacine pouch young. When combined with the appearance of the external features of the thylacine pouch young (Boardman 1945 cited in Sleightholme *et al.* 2012), the histological appearance of the gastrointestinal tract suggests the pouch young were being supplied with milk from the early lactation phase, near the start of the switch phase (Stannard *et al.* 2020). The pouch young thylacines had therefore not yet detached from the maternal teat and were thus fully reliant on their mother for milk, as Boardman (1945, cited in Sleightholme *et al.* 2012) has previously suggested.

Despite the thylacine being extinct, the specimens that remain can still provide some information about the biology of the species. In the future with advances in many different areas of science, such as genomics, and others not yet available, it is hoped we can uncover even more about the thylacine.

Deciphering the processes underlying skull convergence between the thylacine and wolf

Axel Newton

Thylacinus cynocephalus ('dog-headed pouched-dog') was the name given to the large, striped carnivorous marsupial that once thrived on the Australian island state of Tasmania.

Resembling a large canine, the thylacine is in fact more closely related to a kangaroo than a dog. However, through a phenomenon known as convergent evolution, the thylacine and canines, particularly the grey wolf (*Canis lupus*), evolved remarkable similarities in not only their general appearance, but also their skull shape. These similar characteristics are thought to have independently evolved due to adaptation towards similar environmental niches (Wroe and Milne 2007), particularly roles as apex predators that require strong skulls and jaws for taking down prey (Wroe *et al.* 2007; Attard *et al.* 2011).

The similarities in skull shape between the thylacine and grey wolf have not gone unnoticed, being observed and documented for over 30 years (Werdelin 1986). However, only recently, with the advancement of high-powered computing and scanning technology, have detailed analyses into their similarities been performed. A new field of biology, termed 3D geometric morphometrics, has revolutionised anatomy, palaeontology and evolution. Computed tomography (CT) scanning (similar to a medical CT scan) can generate high-resolution, digital reconstructions of complex shapes, such as skulls and bones, and geometric morphometrics allows detailed statistical comparisons between these.

Using these techniques, we examined how similar the thylacine and wolf are by digitally comparing their adult skull shapes (Newton *et al.* 2018). Interestingly, we found that not only do the adult thylacine and wolf skull share superficial similarity, but they also share greater statistical similarity with each other than the thylacine does with its own closest living relatives, such as the Tasmanian devil or eastern quoll (Feigin *et al.* 2018). This was a truly unexpected finding, given that the thylacine and wolf last shared a common ancestor over 160 million years ago, back when dinosaurs still roamed the ancient world.

As evolutionary biologists we want to understand how these remarkable similarities between the thylacine and wolf developed and evolved. Are the instructions to generate their similarities coded in their DNA (see pp. 170–2)? Also, when during their lifecycle did the thylacine and wolf start to exhibit their strong similarities? Were these apparent as pups and joeys or did they arise later as juveniles, during the transition to independent life hunting their own prey? Unfortunately, our ability to answer these questions is hampered by one major obstacle: the thylacine is extinct. However, the tragic tale of the thylacine, hunted to extinction through an imposed government bounty scheme among other threats, also presents an opportunity: a second chance to learn more about the biology of this remarkable species.

When thylacines were captured in the wild, occasionally female animals were found to be carrying joeys in their pouches. In some cases, these joeys were preserved in ethanol fixative and donated to museum collections, where they lay undisturbed for nearly 100 years. Such conditions exquisitely preserved the joeys, essentially freezing them in time. Through collaborations with various Australian and international museums, we sourced all the known thylacine joeys, which were at different ages and stages of development. Using non-invasive CT scanning we digitally reconstructed the internal and external morphology of the thylacine during various stages of its life in the pouch (Newton *et al.* 2018), establishing the first ever digital growth series of an extinct animal.

With this new resource we were able to determine when the thylacine and wolf established their similarities during development. To examine this, we generated an additional digital growth series of the grey wolf from birth to adulthood, allowing us to statistically compare the skulls of the thylacine and wolf in their newborn, juvenile and adult life stages. Staggeringly, we found that not only did the thylacine resemble the wolf as adults, but were also extraordinarily similar as both juveniles and newborns (Plate 3), well before the functional requirements for hunting and taking down prey (Newton *et al.* 2021). Furthermore, we found that their similarities were predominantly observed between bones that develop from a population of embryonic stem cells called the neural crest, cells that are essential for skull development.

Remarkably, these observations complement our other findings that the thylacine and wolf have evolved similar instructions in their genome that influence neural crest stem cells during development (Feigin *et al.* 2019; see pp. 42–4). Together, these findings highlight that the thylacine and wolf have evolved a similar genetic blueprint to generate their similarities in skull shape from the earliest stages of development.

Digital advancements in biological research have allowed us to revisit a classic example of convergent evolution through a modern lens, revealing more about the processes that contribute to evolution and animal diversity. Though hunted to extinction, the fortunate collection and curation of rare thylacine specimens have allowed us a second chance to learn more about this incredible species and how it evolved to resemble a wolf (and vice versa). However, to better learn about the incredible animal diversity on Earth, future studies should not rely on the preservation of specimens in jars. Rather, efforts should focus on the preservation of natural habitats and ecosystems, which will allow direct observations of that animal diversity far into the future.

The brain and behaviour of the thylacine

Gregory S. Berns and Ken W. S. Ashwell

Although several thylacines had been kept in captivity in the early 1900s, no systematic investigation of the thylacine's behaviour was ever documented. The only records of behaviour in their natural habitat are stories passed on by farmers, hunters and trappers (Paddle 2000). Thus, very little is known about thylacine behaviour (Wemmer 2002). Outwardly, the thylacine appears to be an example of convergent evolution, filling an ecological niche similar to members of the canid family elsewhere in the world. As a marsupial, however, the thylacine would likely have brain organisation different from a dog or a dingo.

To further understand thylacine behaviour and to place the thylacine in its evolutionary context, we can look to brain morphology (Berns and Ashwell 2017). By comparing the thylacine's brain structure to that of the Tasmanian devil, we can then infer structural–functional relationships between brain and behaviour. Only four thylacine brains are documented to have survived intact (Sleightholme and Ayliffe 2017). Using magnetic resonance imaging we reconstructed the white matter pathways of two thylacine brains (one on loan from the Australian Museum, the other from the Smithsonian Institution) and compared them with the brains of two Tasmanian devils (one from the Smithsonian, the other from the Save the Tasmanian Devil Program).

The age of the specimens placed limitations on the fidelity of information we could extract; 100 years in preservative takes a toll and was particularly striking in the Smithsonian thylacine. Because of excellent record keeping, we know the original weight of the brain was 43 g. Prior to scanning it was 16 g. Although we don't know the history of storage and preservative changes, we can estimate that the specimen had shrunk at the rate of 1%/year. At such a rate, after 110 years, the specimen would weigh 33% of the original weight.

Despite these limitations, the cortical maps were broadly consistent with both electrophysiological recordings and tract tracing studies in other marsupials and monotremes (Abbie 1940; Karlen and Krubitzer 2007). We found that the motor system was located on the lateral surface of the rostral half of the cortex. Similarly, the somatosensory region was located lateral and ventrocaudal to the motor regions in all the specimens.

Comparing the thylacine with the devil, a few differences were apparent. The motor and sensory regions in the thylacine appeared to be more separated than in the devils' brains, which is be consistent with theories of brain evolution suggesting that cortical fields become more modularised as the cerebral cortex gets bigger (Krubitzer 2007). Other regions of the thylacine cortex appeared larger than the devil's, notably in the frontal regions associated with complex cognition. This would be consistent with the thylacine's ecological niche as a predator, which would require more planning than the scavenging strategy of the devil.

Studying dental development in an extinct marsupial

Tony Harper[1]

Developmental research with extinct organisms is subject to many, fairly severe, practical constraints relative to more typical laboratory studies. Nevertheless, because of its importance for the wider understanding of marsupial evolution and development, a study by Luckett *et al.* (2019) is an attempt to reconstruct the complete dental developmental series for the thylacine (*Thylacinus cynocephalus*) within the context of a comparative sample of other dasyuromorphs (a large group of marsupial carnivores such as the Tasmanian devil, quolls, dunnarts, antechinuses and the numbat).

A principal focus of this study was the description of histological preparations made from the pouch young of an adult female thylacine collected in Tasmania in the winter of 1909 and accessioned at The Museums Victoria (NMV). This approximately 32-day-old litter included the male specimen (NMV C5754; see also Old 2015) chosen for histological analysis and his sister (NMV C5757), which was sampled for the whole-genome analysis of Feigin *et al.* (2018). The comparative and histological information summarised by Luckett *et al.* (2019)

1 Eds: This chapter is based upon a paper written by W. Patrick Luckett (deceased), Nancy Hong Luckett (deceased) and Tony Harper (Luckett *et al.* 2019). Unfortunately, it was not possible to contact any family members of the deceased in order to seek permission to print their names alongside that of Tony Harper.

is therefore intended to update our knowledge of thylacine biology and to emphasise the continuing importance of research on the dentitions of marsupials generally.

Scientific attention to the developing dentition of the (then extant) thylacine began with the comprehensive work on marsupials by William Henry Flower in 1867 (Flower 1867). At that time Flower considered the thylacine's dental replacement pattern to represent a 'rudimentary diphyodont' condition, in accord with then commonplace assumptions about the adaptive inferiority of marsupials relative to their placental mammal counterparts. Although more recent studies have described, and even emphasised, the derived nature of the ancestral marsupial dental formula and replacement pattern relative to contemporaneous placental mammals; the wider mammalogical community has been slow to adopt these newer perspectives. For instance, studies affirming the identity of the first two upper and lower postcanine teeth in adult marsupials as deciduous premolar 1 (abbreviated as 'dP1') and deciduous premolar 2 ('dP2'; Luckett 1993; Luckett and Woolley 1996) have not curbed the common practice of referring to these teeth as the successional 'P1' and 'P2' in many publications on marsupials.

For its part, the literature on marsupial dental ontogeny is often intimidating to non-specialists because of the large amount of embryological detail required to locate and identify minute and often poorly differentiated vestigial tooth germs and dental laminae. In marsupials, the only postcanine tooth position to undergo a 'normal' (from a placental mammal's perspective) pattern of deciduous and then permanent tooth eruption is the third (and last) premolar in both upper and lower jaws. However, relative to the location of the shedding deciduous third premolar, the successional third premolar often erupts a significant distance mesio-lingual to its first-generation precursor. This often causes the deciduous third premolar's 'lingual successional lamina' (a ribbon-like epithelial cord connecting the developing follicle of dP3 to its replacement P3) to take a very tenuous and anteriorly directed course, paralleling the epithelium lining the oral cavity.

Because this thin and superficial lingual successional lamina can be difficult to locate at a distance from the dP3 follicle, and is easily misdiagnosed as a direct invagination of the oral epithelium, Luckett *et al.* (2019) emphasise the importance of analysing and illustrating a dense and complete sample of histological cross-sections when making identifications of marsupial developmental structures. The illustration in Fig. 2 is of an oblique histological preparation of the developing mandible of NMV C5754 (drawn by Nancy Hong-Luckett), which demonstrates the connectivity between the dP3 and P3 follicles through their deteriorating lingual successional lamina, and highlights the importance of detailed anatomical observations for the correct diagnosis of embryonic structures.

The analysis of between-species differences in dental development and eruption patterns among extant marsupials is also complicated by the wide range of body sizes and developmental rates observed among the pouch young of these species. The descriptions in Luckett *et al.* (2019) of the relative maturation of P3 seen in *Thylacinus* versus a representative sample of extant dasyurids relied on contrasts between the ontogenetic stage attained in the upper and lower P3 versus the corresponding second molar (M2). Using this standard, Luckett *et al.* (2019) found that two basic postcanine developmental patterns are apparent among the dasyuromorphians sampled. The extant two-premolared dasyurids (the daysurines *Sarcophilus*

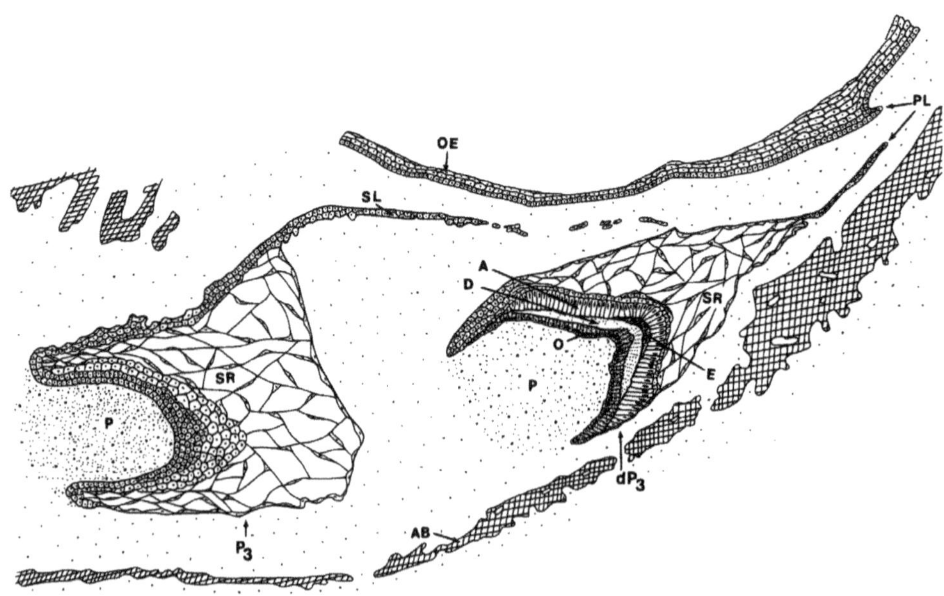

Fig. 2. Illustration of a histologic cross-section through the lower jaw of a thylacine pouch young. The developing follicles of the deciduous lower third premolar (dP3) and its replacement successional third premolar (P3) can be clearly seen, as can the delicate epithelial cord connecting them – the lingual successional lamina (SL). In extant two-premolared dasyurids detailed histological preparations such as this are critical for identifying the remaining premolar positions. For example, in many cross-sections the SL can be mistaken as an independent invagination from the lining of the oral epithelium (OE), thus causing the P3 to be confused with a deciduous premolar. Other abbreviations include: A, ameloblasts; AB, alveolar bone; D, dentin; E, enamel; O, odontoblasts; P, dental papilla; PL, primary dental lamina; SR, stellate reticulum. Reprinted with permission from Luckett, WP, Hong-Luckett, N, Harper, T (2019). Microscopic analysis of the developing dentition in the pouch young of the extinct marsupial *Thylacinus cynocephalus*, with an assessment of other developmental stages and eruption. *Memoirs of Museum Victoria* **78**: 1–21.

and *Dasyurus*; which are missing dP2) and three-premolared *Thylacinus* show an accelerated development of P3 (relative to M2) and a concomitantly smaller and minimally functional dP3. Conversely, the extant three-premolared dasyurids studied (the dasyurine *Antechinus*, and sminthopsine *Sminthopsis*) show a retarded maturation and eruption of the successional P3 (relative to M2) and therefore a relatively longer functional lifetime of dP3 in the juvenile jaw.

These results have some perplexing implications for the evolution of dental formulae and replacement patterns near the evolutionary origin of the order Dasyuromorphia. Whereas the modern dasyurids, which retain an ancestral postcanine dental formula (three premolars and four molars), show a delayed development of the successional P3 (similar to the condition in non-dasyuromorphian taxa such as *Perameles* and *Monodelphis*), the modern two-premolared dasyurids studied show greater similarity to the thylacine (and likely also *Myrmecobius*) in having a quickly developing and erupting successional P3. This accelerated loss and replacement of dP3 is more easily explained in the two-premolared *Sarcophilus* and *Dasyurus*, given the limited space available for the eruption of new teeth within the relatively compacted postcanine tooth rows found in these taxa. However, this is definitely not the case in the thylacine, which retains the ancestral complement of three premolars within a relatively elongate jaw. Later adult developmental stages of *Thylacinus* even show a widening of the spaces separating consecutive premolars (termed 'diastemata'), making it unclear why the rapid loss of dP3 would be selectively advantageous.

The study by Luckett *et al.* (2019) therefore raises many questions about whether the heterochronic variation in dP3 replacement can be attributed to evolutionary adaptation or ancestral baggage in *Thylacinus* and the other dasyuromorphians sampled. We sincerely hope that this work will stimulate the curiosity of newer generations of embryologists interested in the amazing life histories of the thylacine and other dasyuromorphians.

Thylacine: the skeleton of a cursorial marsupial

Natalie M. Warburton

The thylacine skeleton, like that of all animals, functions to protect vital organs and support the weight of the body and works together with muscles to produce controlled movements of the body. The shapes of the various bones reflect their interactions with the muscular system and the mechanical forces that act through the body as it moves within the environment (Warburton and Dawson 2015). Skulls and teeth provide clues as to how animals find, catch and process food, while limbs (appendages) can tell us about how an animal moves. Other aspects of the skeleton hold evidence of an animal's ancestry. For example, characteristics such as the presence of epipubic bones or the absence of a bony patella unite marsupials as a group.

The thylacine skeleton suggests that, in comparison with all other marsupials, thylacines were well-adapted for running quickly and efficiently (cursorial adaptation), presumably in order to catch prey (Warburton *et al.* 2019). Long, thin limb bones (humerus, radius, ulna, femur, tibia, fibula) enable longer strides, such that distance can be covered more efficiently than with short limbs. The muscle attachments on the long bones correspondingly reflect a relatively 'high-geared' arrangement, whereby muscles are able to achieve a distance or speed advantage. Long in-levers such as the calcaneus (heel bone), similarly suggest investment for rapid limb movements in line with the body axis rather than modifications for high-force output from the limbs.

Limb posture also provides important clues for understanding adaptation for locomotion. The carpal (wrist) and tarsal (ankle) bones of thylacines are relatively deep and narrow and support a more upright limb posture standing on the toes (digitigrade), rather than a flat-footed plantigrade stance. Correlated with this, the metacarpal (hand) and particularly metatarsal (foot) bones are long and thin and contribute extra length to the limb in a digitigrade stance, which functions to further increase stride length and thus contribute to more efficient locomotion at high speed.

Efficient running locomotion also requires modification of the attachment between the limbs and the trunk via the limb girdles. The forelimb is primarily attached to the trunk by muscles that attach to the scapula and humerus, with the clavicle providing stability and support for the position of the shoulder joint. Thylacines, however, had a very reduced clavicle, which enabled the forelimb to move more easily in line with the body for rapid running locomotion. Similar adaptations are found in running placental mammals.

As mentioned before, marsupials generally possess an elongate epipubic bone that is connected to the front of the pelvis and the muscles of the abdominal wall. This bone is also found in monotremes and reptiles but has been lost in the evolution of eutherian ('placental') mammals. It has been suggested that epipubic bones may help to stabilise the trunk during locomotion. Thylacines, however, had very reduced epipubic bones in comparison with all other marsupials (except the very unusual marsupial moles), which suggests that thylacines evolved quite different adaptations to locomotion than other marsupials, with the reduced epipubic bones presumably allowing greater flexibility of the trunk for rapid running locomotion.

Collectively, these features demonstrate remarkable specialisation in the thylacine skeleton for cursorial running locomotion that, evolutionarily, is convergent with other running mammals. There is still much to be learnt about the evolution and development of the thylacine skeleton though. It would be interesting to examine the rate of growth of the different parts of the skeleton and also to examine the early development of features such as the reduced clavicle and epipubic bones, so that we can better understand the forces that influenced the evolution of this unique and beautiful marsupial.

How thylacines walked

Matt Cartmill, Christopher Atkinson, Kaye Brown, Erica A. Cartmill, Daniel Gonzalez-Socoloske and Adam Hartstone-Rose

Animals with legs use them to propel themselves, often in irregular bursts of movement. When they repeat a regular pattern of propulsive limb movements over and over at a constant speed, the pattern is called a 'gait'. Gaits are either symmetrical, in which the pattern in the second half of each cycle is a mirror image of the first half (with left and right reversed, as in a human walking: left, right, left, right), or asymmetrical (e.g. a kangaroo hopping with both legs moving together). As an animal walks faster and faster, each foot stays planted for a smaller percentage of the cycle, until there is a part of the cycle ('aerial phase') when the animal is flying through the air with all feet off the ground. At that point, the animal is no longer walking, but running.

Any symmetrical four-footed gait can be specified by just two numbers: *duty factor* (the percentage of time that a foot is planted on the ground, which is usually about the same for all four feet and varies roughly as the inverse of speed) and *diagonality* (the phase difference between fore- and hindlimbs: the percentage of total-cycle duration by which the forelimbs repeating left-right cycle lags that of the hindlimbs). In a symmetrical gait with very low or very high diagonality, the animal puts both left feet down together and then both right feet,

Fig. 3. Eight frames representing the eight footfalls (F = fore, H = hind, L = left, R = right) from two successive strides of the thylacine 'Benjamin'. Arrows indicate the foot that has just touched down in each frame. The superimposed pawprints, added to mark the position of each footfall, present a schematic curved trackway (bottom). In these cycles, the animal's footfalls are almost evenly spaced in time, giving it a diagonality of approximately 25 and placing it near the boundary between diagonal and lateral couplets (see Fig. 4). Film stills courtesy of The Thylacine Museum <http://www.naturalworlds.org/thylacine/>.

like a camel walking ('lateral couplets'). In a gait with moderate diagonality, limbs that are diagonally opposite move together, as in a trotting horse ('diagonal couplets').

Different mammals walk symmetrically in different ways, putting their feet down in footfall sequences that vary among orders and families. The few existing films of captive thylacines give us an opportunity to see which of the other mammals the thylacine most resembles. Thylacines resemble the dog family (Canidae) of placental mammals in diet, dentition and body form and so we might expect them to have dog-like locomotion as well. We have previously described two thylacine walking cycles extracted from one of the films which resembled the slowest walks of dogs (Cartmill *et al.* 2020). The recent discovery of new films of caged thylacines provided an opportunity to increase our sample.

We studied all nine known films of thylacines and identified potentially usable gait cycles from four of them.[2] These cycles were converted into still frames using the 'Render Video' function in Adobe Photoshop and the frame numbers for each footfall and liftoff were determined for calculating duty factor and diagonality (Fig. 3). Because these films were not made with gait analysis in mind, image resolution is low and the camera angles are less than ideal, so different observers sometimes had different estimates of the exact frame in which a foot was lifted or put down. To help correct for interobserver error, we ran four independent

2 Eds: this chapter was written before the discovery of the tenth known film (see pp. 111–12), which unfortunately does not contain a useable gait cycle for the juvenile animal.

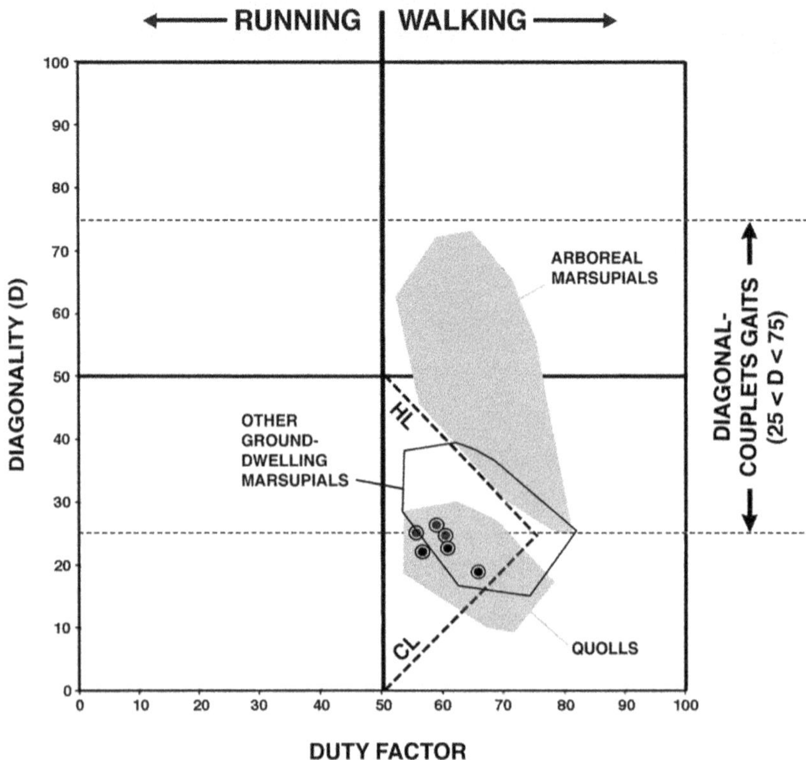

Fig. 4. Estimated duty factor and diagonality in thylacine walking gaits, averaged across four observers for each of six symmetrical walk cycles (circles). Diagonal dashed lines represent the theoretical 'horse line' (HL; see text) and 'camel line' (CL). Lower grey area: spotted-tailed quoll (*Dasyurus maculatus*). Solid outline: other ground-dwelling marsupials. Upper grey area: arboreal marsupials. Non-thylacine data from (Cartmill *et al.* 2020).

analyses of each of the cycles examined and came up with just six gait cycles determined by all four analysts to be symmetrical. These six cycles are from two films, both of the same individual ('Benjamin') in the Beaumaris Zoo in Hobart (Sleightholme *et al.* 2020).

The values of duty factor and diagonality reported by the four observers were averaged for each of these six cycles and plotted on a 'Hildebrand diagram', a graph of duty factor against diagonality (Fig. 4). Our analyses showed that the surviving records of thylacine locomotion all represent moderately slow walks, clustered around the boundary between lateral-couplets and diagonal-couplets gaits, similar to the recorded gaits of quolls (*Dasyurus* spp.) in captivity. Like the walks of most other terrestrial marsupials, the observed thylacine gaits all lie below the so-called 'horse line' (Fig. 4). Gaits in this part of the Hildebrand diagram minimise interference between fore- and hindfeet in walking (Cartmill *et al.* 2020).

We predict that thylacines, like other terrestrial marsupials, would have exhibited higher diagonality values in walking as speed increased and duty factors went down (following the horse line upward), but would not have displayed the behavioural flexibility seen in the gait-pattern distributions of dogs (i.e. encompassing both horse-type and camel-type gaits:

Cartmill *et al.* 2002). We will probably never know for sure unless living thylacines are discovered or new films come to light showing thylacines moving around at more varied speeds in less cramped enclosures. However, our prediction should also hold for Tasmanian devils (*Sarcophilus harrisii*), the thylacine's largest surviving relative – which, happily, are still around to be studied.

Acknowledgements

We are grateful to The Thylacine Museum's Curator, Cameron Campbell, for his generous provision of all the known films of living thylacines and for his kind permission to use the stills reproduced in Fig. 3.

The likely hunting behaviour of the thylacine, as deduced from its forelimb anatomy

Christine M. Janis and Borja Figueirido

The thylacine has been popularly called the 'marsupial wolf', especially in the Northern Hemisphere, and is often used as a classical example of convergent evolution between marsupials and placentals. But despite its superficially dog-like appearance, the thylacine was certainly not an ecological analogue of the placental wolf. Rare observations from the early 20th century report it hunting alone or in pairs, with neither the pursuit chasing of a wolf nor the specialised ambush attack of a large cat (Smith 1982). Were thylacines perhaps like other, more generalised canids, such as coyotes or jackals? These animals hunt via a type of behaviour called 'pounce–pursuit', which involves a short chase before pouncing on the prey.

Could the skeleton of the thylacine inform us as to its predatory behaviour? To answer this question, we must determine whether such behaviours correlate with the anatomical features of living animals with known behaviour. Scientists have been making investigations into the correlation between anatomy and behaviour (and/or performance) in animals for many years. In the past couple of decades techniques have become increasingly sophisticated, including the use of computed tomography scans of bones to image the anatomy in detail, and a diversity of statistical techniques to analyse the data collected.

Some previous studies on general limb proportions (Jones and Stoddard 1998; Jones 2003) showed the thylacine to be a generalised carnivore, less specialised for either running or pouncing than living placental carnivores and more similar to its marsupial relatives, the quolls and the Tasmanian devil. We decided to look specifically at the forelimb, because that reflects whether a carnivore is specialised for running (e.g. wolves) or retains the ability to grapple with its prey (e.g. cats).

A particular feature that we noted was the structure of the elbow joint, specifically the anatomy of the lower end of the humerus (the upper arm bone). When humans rotate a hand from the palm facing down position (prone) to the palm facing up (supine), the movement comes from the rotation of the forearm (comprising the radius and ulna bones) at the elbow joint. When pronating the hand, the radius rotates over the ulna: this ability to swivel the forearm bones is reflected in the anatomical articulation on the humerus.

Most mammals have less ability for forearm rotation than humans: their forelimbs are more restricted to a prone, palm down position. In specialised running mammals such rotation becomes even more restricted. Compare throwing a ball of wool to a cat versus a dog. The cat retains the more generalised mammalian ability and will try to catch it in its front paws, turning the paws inwards to a certain extent, whereas the dog will rather try to pat the ball with the paw facing the ground; dogs have become more specialised in their elbow anatomy. Dogs can manipulate objects with their paws to a limited extent (e.g. a holding a bone to chew on it), but a horse cannot rotate its hooves like this at all. In general, cursorial (running-adapted) carnivores have restricted motion in their elbow joints.

Where does the thylacine fit on this spectrum of forelimb mobility? Was it more like a specialised dog or more like other mammals including cats? We compared the shape of the humerus of eight different thylacine individuals with those of a diversity of living carnivorous mammals (Figueirido and Janis 2011): all of the thylacines clustered with members of the cat family, showing the more generalised mammalian condition. Not only were they distinctly different to wolves, they were also different to the smaller pouncing canids such as coyotes and jackals.

In a broader comparison of all of the forelimb bones with those of diverse carnivorous mammals (Janis and Figueirido 2014), the thylacine was still generalised in its anatomy. It lacked the specialisations of both the pursuit and pouncing dogs on the one hand and the ambushing cats on the other, but neither did it ally with its carnivorous marsupial relatives such as Tasmanian devils and quolls. Thylacines appear to have been more generalised as medium-sized carnivores than their placental counterparts elsewhere in the world, perhaps because they lacked competition during their evolutionary history in Australia (until possibly the arrival of dingoes). Quolls have always been small generalists, Tasmanian devils are bone-crushing scavengers and the now-extinct marsupial lions (family Thylacoleonidae) were apparently highly specialised ambush predators, the smaller ones being arboreal.

Thylacines were medium-sized, terrestrial, somewhat cursorial carnivores (more so than other marsupials), but not as specialised for running as placental cursors. They took this ecological role in Australia without apparent competitive pressure to become more specialised. In contrast with the evolutionary histories of the families of placental carnivores, where migrations and immigrations mingled faunas and resulted in competition, thylacines evolved in relative isolation and so remained generalised.

Weighty implications of the thylacine's body mass

Douglass S. Rovinsky

One of the most basic and important aspects of an animal is its body mass. An animal's body mass directly affects a myriad attributes, including metabolic rate, lifespan, growth and reproduction, what it can (and has to) eat, interactions with the environment, habitat and what role it plays in the ecosystem. This makes knowing the average body mass of a species incredibly important, especially if that animal is extinct. When reconstructing the animal's functional ecology – how it lived and interacted within the ecosystem – body mass is a crucial piece of the puzzle.

Surprisingly, we have a poor understanding of the body mass of the average thylacine, despite the taxidermy and wet-tissue specimens, photographs and video footage, because the species was driven to extinction before it was common to record a specimen's weight. In fact, there are only four reliable records of the weight of individual thylacines: two in the published scientific literature (both ~15 kg; Crisp 1855; Berns and Ashwell 2017) and two recorded in the Register of Deaths of the London Zoo (13.2 and 26.1 kg, respectively). Of all the thylacines caught, exhibited, photographed or filmed, these four well-documented body masses are all that are known to exist.

Other than those four records, there are a few anecdotal accounts scattered throughout 19th century Tasmanian periodicals. These historical documents formed the basis for the 29.5 kg average mass estimate commonly used for the past two decades (Paddle 2000). However, these early reports all share two similarities: they record wildly different weights for sometimes similarly sized thylacines (27.2–52.6 kg), and all are from after 1830, the start of the Van Diemen's Land Company bounty scheme. There is a single known report of a thylacine's weight from before it became lucrative to kill them – the Lt Governor Paterson account from 1805 (Paterson 1805b). This account records a weight of 20.4 kg for a large adult male, much smaller than the 29.5 kg population average given by Paddle (2000) or the accounts from after the initiation of a bounty scheme (average of 34.5 kg).

Recent quantitative analyses of surviving thylacine specimens suggest a much smaller population average of 16.7 kg (Rovinsky *et al.* 2020). That study found that females would have commonly ranged between 11 and 17 kg, and males between 15 and 25 kg. These estimates closely match the four reliable mass records and, interestingly, match the early 19th century, pre-bounty scheme Paterson (1805b) record as well. This reduction in the thylacine's body mass has profound implications for our understanding of its biology and functional ecology.

Thylacine males were approximately 44% heavier than females, indicating that they would have been functionally sexually dimorphic (≥20% difference in body mass; Ruckstuhl and Neuhaus 2002). This level of dimorphism is very close to that seen in the chuditch (*Dasyurus geoffroii*, 47%; Serena and Soderquist 1989). Sexual dimorphism in marsupials correlates with biology, life history and social attributes (MacFarlane *et al.* 2005), allowing some grounded speculation on aspects of the thylacine's ecology.

As in the chuditch, female thylacines were probably monoestrus, with a relatively synchronised mating season spanning 2–8 months (potentially centred around December; Dixon 1989; MacFarlane *et al.* 2005). Female thylacines likely held non-overlapping territories, which they would defend from other females, with the potential exception of their young or breeding daughters (Serena and Soderquist 1989). Reports of 'packs' of thylacines were almost certainly female thylacines and their juveniles, representing home territory tolerance of young prior to dispersal (Serena and Soderquist 1989). Males would have held larger, looser territories overlapping with that of several females and other males. There was probably a degree of intrasexual (within sex) aggression between males, especially during the mating season; for example, one study found that nearly 50% of chuditch males bore wounds resulting from male–male combat, compared with only 5% of females, with at least 83% of these wounds occurring during the mating season (Serena and Soderquist 1989). This intolerance by males of non-mating conspecifics is also suggested by anecdotal reports, which rarely report adult males seen with a female and her young (Dixon 1989).

This size dimorphism also suggests that males may have had a feeding ecology different from females, as seen in the sexually dimorphic quolls (*Dasyurus* spp.; MacFarlane *et al.* 2005). Mammalian carnivores under 21 kg in mass tend to prey on items smaller than 45% of their own mass, due to a cost–benefit energetics requirement (Rovinsky *et al.* 2020). This cost–benefit threshold suggests that female thylacines would have normally preferred prey weighing 5–8 kg, and males 7–12 kg. Male thylacines' greater mass would have let them take nearly all available Tasmanian prey species, including the red-necked wallaby (*Notamacropus rufogriseus*) or even small/subadult grey kangaroos (*Macropus giganteus*) and common wombats (*Vombatus ursinus*). Females and subadult males would likely have preferentially targeted the smaller-bodied bilbies, bandicoots and pademelons. Although this does seem to exclude the thylacine from an 'apex predator' niche, we should remember that the extinction of the widespread, jaguar-sized *Thylacoleo carnifex* occurred only within the past 40 000–50 000 years. This large 'marsupial lion' would undoubtedly have occupied this apex predator niche until recent prehistory.

The persecution of the thylacine was based on the concept of it being a large, formidable predator, hunting and killing colonists' livestock with reckless abandon. Stories abound of this nature – including some that mention almost supernatural strength, near-invulnerability and a rapacious hunger. These stories, much like the inflated reports of their size, may have been knock-on effects or even encouraged by the bounty culture surrounding the thylacine – giving a reason to exterminate and adding more fuel to the fire. By returning to the thylacine specimens themselves, we can discover evidence-based refutations of many of these claims. We are now letting the animal tell its own tale, albeit too late.

Were thylacines wrongly persecuted? Truth behind the jaws

Marie R. G. Attard and Stephen Wroe

Around 300 sheep were introduced into Tasmania in 1803 and their numbers rapidly increased to 172 000 by 1819 (Paddle 2000). A very small population of thylacines (*Thylacinus cynocephalus*), estimated between 2000 and 4000 individuals, persisted when Europeans first settled there (Prowse *et al.* 2013). Research suggests that most thylacines were unlikely to have been in contact with sheep (Prowse *et al.* 2013) and those that were killed as part of the government bounty scheme (1888–1909) were primarily from areas outside of grazing lands (Davies 1965). Even if their home range overlapped agricultural areas and given that they were likely solitary predators, one must ask: 'Were thylacines physically capable of taking down a full-grown sheep?' To answer this question we need a better understanding of the maximum prey size constraints of the species.

The thylacine's dentition points towards a hypercarnivorous diet, consisting primarily of meat (Wroe *et al.* 2008; Figueirido and Janis 2011). They were unlikely to be regular scavengers or bone consumers based on their dental features (Jones and Stoddart 1998). The prey size limit of thylacines has been highly contested over recent decades due to conflicting scientific evidence and anecdotal accounts. There are numerous small–medium-sized native species (<1–5 kg) that they almost certainly would have targeted based on bone deposits discovered in caves (Case 1985) thought to be used as dens or resting places (Guiler 1985). These include bettongs, ringtail possums, potoroos, bandicoots and small species of *Macropus*. However, records from such bone deposits may give an incomplete picture as it is possible that larger prey were not carried back to the den.

By measuring the ratios of different isotopes in animal tissues researchers can trace the behaviour of that individual animal, such as the diet and the environment. Stable isotope signatures taken from hair and bone of 34 preserved thylacines suggest they hunted a variety of prey species (Attard 2013). The stable isotope values of their tissues overlap with other co-occurring marsupial carnivores in Tasmania, raising the possibility that these top predators competed for the same prey species as Tasmanian devils (*Sarcophilus harrisii*) and spotted-tailed quolls (*Dasyurus maculatus*).

It has been thought that thylacines had a relatively large gut (Crisp 1855) to achieve distinct reductions in hunting activity (Guiler 1985). Possibly their kill frequency could have been as low as once every 3 days, though it is difficult to know for certain. Tasmanian devils and quolls have a metabolic rate comparable to that of non-carnivorous marsupials of a similar size, allowing them to conserve energy and hunt both small and large prey. This differs from placental carnivores, which have higher basal metabolic rates. The metabolic rate of the thylacine was likely similar to that of other large carnivorous marsupials, helping these animals to survive unfavourable conditions and/or conserve energy.

Mammalian carnivores have specialised head and jaw anatomy for capturing, handling and killing prey. Bite force and the skull and tooth morphology can be used together to assess the ability of a species to take larger or more difficult-to-handle prey. Thylacine skulls have a long snout that would enable them to bite down at high speed, a characteristic of predators specialising in relatively small, more agile prey and those that use a pounce–pursuit or ambush hunting style. Some have claimed that the thylacine's jaws could open to as wide as 120° to grab large prey (Mittelbach and Crewdson 2005), but more realistically could open between 75° and 80°, which is considered a wide gape for a mammal. Wide gape size may have allowed them to catch prey of a variety of sizes, including larger species (Johnson and Wroe 2003), or increase bite depth when killing prey. Alternatively, it is possible that thylacines used their wide gape as a threat display, as do Tasmanian devils. Mammalian carnivores have a greater risk of fracturing or breaking their teeth from catching and killing larger prey or eating bone than species specialising in smaller prey, because, as they struggle, larger prey produce relatively high, unpredictable loads on the predator's jaws. Low rates of canine tooth wear and fracture suggest that thylacines relied on prey ranging from <1 to 5 kg (Jones and Stoddart 1998).

Over the past 15 years or so 3D computer simulation has been increasingly used to model mechanical performance in extinct, as well as living species. This approach can predict how an animal's skull might respond to the forces generated in the capture and incapacitation of live prey. They are then typically compared to their closest living relatives to evaluate the relationship between a species' diet and skull biomechanics (Attard *et al.* 2011). These digital models are created from computed tomography scans, allowing the precise geometry and different bone densities of the skull to be incorporated. Muscle forces can be applied to simulate the jaws closing on prey and the resulting patterns and magnitudes of stress and strain in the skull can then be predicted.

Mammalian carnivores typically bite down then shake their head to restrain or kill prey. These behaviours can be also simulated to examine the mechanical response of the skull to these loads. In thylacines, simulations show that the front of the snout would have experienced extremely high stress when the animal shook prey from side to side (Attard *et al.* 2011), which may have put a mechanical limit on prey size. Other marsupial carnivores inhabiting the same geographical area, such as the Tasmanian devil and spotted-tailed quoll, are better adapted to capture relatively large prey when the same types of loads are applied to the skull. Similarly, Dickson's thylacine (*Nimbacinus dicksoni*) – an extinct fox-sized thylacinid – had a high bite force for its size, comparable to that of Tasmanian devils and larger quolls known to take down prey several times heavier than themselves (Attard *et al.* 2014).

Several computer-based studies have developed 3D models of adult thylacine skulls to study the impact of prey capture and their effects on skull structures. Wroe *et al.* (2007) constructed computational models of a thylacine and dingo (*Canis lupus dingo*) skull to predict which species was better adapted to catch and kill large prey. Although simulations predicted that the bite force of the thylacine would be higher than that of the dingo, the dingo's skull was far better adapted to sustain these loads, suggesting that they had the ability to outcompete thylacines for prey where their ranges overlapped.

The mechanical weakness of their long narrow snout (Jones 2003) presents a morphological disadvantage that would have limited prey size selection by thylacines, especially if they hunted alone. When hunting large, difficult-to-catch prey, mammalian predators may selectively hunt weaker individuals. Other mammalian predators can kill larger prey when hunting communally and thereby counteract morphological constraints of the jaw. Historical accounts describe thylacines hunting alone, in pairs or small family groups of up to two adults and four young. However, population viability analysis models suggest their population size was small prior to European habitation (Prowse *et al.* 2013) and was in decline over the time they were observed (Paddle 2000). As such, their small hunting group size may be an artefact of their dwindling population. Still, it is unlikely that thylacines were ever true social hunters such as dingoes.

History consistently shows that people are more fearful and less tolerant of large predators and consequently, they are more likely to be targeted for persecution and exploitation. Thylacines were considered among the most threatening of Tasmanian fauna because of their large, wolf-like appearance. Their persecution was first encouraged by the Van Diemen's Land Company, which placed a bounty on the species in 1830, fuelled by the notion that they were switching to a new prey source – farmer's livestock (Paddle 2000). There is no doubt that at least some thylacines would have killed lambs or possibly even sheep, especially if the prey was in poor condition, but how extensive was this problem?

More recently in Australia, high predation rates of livestock by foxes and other introduced predators only applies to a small area of country or is linked to circumstances peculiar to a single flock. Flocks in closer proximity to optimal predator habitats and in smaller groups have greater predation risk. Lambs of breeds with poor mothering ability, such as the Merino, are also more likely to be picked off. The first sheep to arrive in Tasmania were likely crossbred Bengal and Cape breeds (Kirkpatrick and Bridle 2007), but within several years their bloodlines were mixed with Merinos to improve the quality of their fleece.

Eradication of top predators does not necessarily correspond with reduced loss of livestock. Most recent studies show no effect of fox control on lamb production. However, fox control measures significantly reduced the percentage of lamb carcasses classified as killed by foxes (Greentree *et al.* 2000). The actual cause of lamb fatalities is difficult to estimate, as supposed livestock 'kills' may in fact represent individuals that died natural deaths, later scavenged by the accused predator. Our best modern estimates of thylacine body mass, biomechanical limitations and predatory strategy indicate this species was poorly adapted to handle large-bodied prey, especially if hunting alone, and it seems unlikely that thylacines were regular hunters of sheep.

The thylacine is frequently presented as a cautionary tale of wrongful persecution, which we can now verify with scientific evidence linked to maximum prey tolerance. This is not a stand-alone incident. Hunting of large endemic carnivores still occurs in Africa (lions and leopards), North America (wolves, bears, mountain lions) and Eurasia (wolves, bears, lynx), despite the fact that these species play key ecological roles at the top of the food chain. Regardless of their potential effect on the livelihood of settlers, we are now aware that the loss of any large native predator can have catastrophic knock-on effects down the food chain. Although the ecological imbalance caused by local decimation of predators is reversible, this realisation comes far too late for the thylacine and the ecosystem over which it once ruled.

Menagerie of a ghost: parasites of the Tasmanian tiger

Mackenzie Kwak

Parasitism is the most common mode of existence in the living world, so it should come as no surprise that every free-living animal hosts its own parasitic microcosm: a personal menagerie. As well as their ubiquity, parasites are also ancient. Biologists have discovered dinosaur-eating ticks in Mesozoic amber and parasitic botflies in the stomach of a woolly mammoth frozen in the Arctic tundra. Yet despite their great diversity and abundance, we know very little about the parasites of most extinct species, largely because many biologists in the colonial era have long overlooked them.

In fact, in some cases, the co-extinction of parasites is not realised until decades or even centuries after their passing. For example, the Christmas Island flea (*Xenopsylla nesiotes*) vanished with the Maclear's rat (*Rattus macleari*) in 1903 (Kwak 2018), but this co-extinction was not discovered until over a century later! As the biology of the thylacine was never studied in detail, one might assume that we would know nothing of its parasites. Fortunately, thanks to the careful work of the parasitologists of the past we know a surprising amount about some of the parasites that lived on and in the thylacine.

Parasites are generally grouped into two broad camps: ectoparasites (which live on the outside of their host) and endoparasites (which live on the inside). Only a single ectoparasite species, the burrowing flea (*Uropsylla tasmanica*), has ever been recorded from the thylacine (Dunnet and Mardon 1974), based on a single adult female specimen collected from a thylacine in Launceston in 1879 and a number of endoparasitic larvae found embedded in thylacine pelts in various museums (Pearse 1981). Fortunately, we can glean an understanding of how this ectoparasite would have affected thylacines based on how it affects its extant hosts: quolls (*Dasyurus* spp.) and Tasmanian devils (*Sarcophilus harrisii*).

The burrowing flea is unique among all living fleas because it tunnels beneath the skin of its host and behaves as a subdermal endoparasite in its larval stage (Kwak *et al.* 2017). As an adult, it behaves like any other flea and lives in its host's fur and feeds on blood (Plate 4). It is likely that infestations caused by the larvae of the burrowing flea in the thylacine resulted in localised skin lesions, as is the case in related species, such as quolls (Kwak *et al.* 2017). Fortunately, the burrowing flea is not just a parasite of the thylacine, so when that host disappeared the flea was saved from extinction thanks to its taste for the blood of quolls and Tasmanian devils.

Following the death of a captive thylacine in the National Zoological Park in the USA in 1904, a parasitologist discovered a small immature tapeworm, no more than a few millimetres long, in the animal. The 'Tasmanian wolf tapeworm' was described and given the scientific name *Dithyridium cynocephali* and was largely forgotten about for the next 80 years (Ransom 1905). In 1989, a pair of parasitologists re-examined the sole specimen of the 'Tasmanian wolf tapeworm' and concluded that it was identical to the Tasmanian devil tapeworm (*Anoplotaenia dasyure*) (Obendorf and Smith 1989). So, what was a Tasmanian devil tapeworm doing in a thylacine and how did it get there? To answer that we need to understand the life cycle of the Tasmanian devil tapeworm.

Life for this strange little parasite begins when it is excreted as an egg in the faeces of its host. Most tapeworms need at least one intermediate host to begin their development and this tapeworm's preferred intermediate hosts are the Bennett's wallaby (*Macropus rufogriseus*), Tasmanian pademelon (*Thylogale billardierii*) and long-nosed potoroo (*Potorous tridactylus*) (Gregory *et al.* 1975). Once they enter one of these hosts they develop into their penultimate form: a metacestode. The metacestode is a tiny parasite only a few millimetres long that encysts in the tissue of the intermediate host and waits to be eaten by its final host: a Tasmanian devil or quoll. When the host is consumed, the metacestode makes its way to the host's intestine and begins to grow into its final form that we are familiar with: the slender tapeworm. It now produces eggs to begin the cycle again. It is likely that the thylacine became infected when it consumed the egg-laden faeces of a devil with a tapeworm, or the infected devil itself. The tapeworm eggs then hatched and grew into metacestodes and remained in the tissue of the thylacine.

One other species of parasitic worm has been recorded from the thylacine, though it caused some confusion and minor embarrassment along the way! When visiting the Department of Parasitology at the London School of Hygiene and Tropical Medicine, the great Australian parasitologist John Sprent found a vial of four roundworms labelled as having come from a 'Tasmanian Marsupial Wolf', and must have been collected from a captive thylacine in the London Zoo. Having never seen a parasitic worm like these from an Australian marsupial, Sprent described and named them *Cotylascaris thylacin* (Sprent 1971). However, just a year later, Sprent realised his mistake.

These were not a new parasitic worm at all, but instead the pigeon roundworm (*Ascaridia columbae*), one of the most common parasites on the planet (Sprent 1972). How had they ended up in a thylacine though? Sprent reasoned that the captive thylacine must have caught and eaten a hapless pigeon (*Columba livia*) that had strayed into its enclosure at the zoo. Although the bird had been digested, the worms could withstand the process, but could not live in this new host and so died and were passed out in the animal's faeces, only to be discovered and preserved.

Though we will likely never know the full diversity of the thylacine's parasites, small fragments of data provide a glimpse into the fascinating parasitology of this enigmatic creature (Plate 5). And by studying how these parasites affect the close relatives of the thylacine, we may better understand Australia's most mysterious marsupial.

Thylacine immunogenetics and de-extinction

Emma Peel, Carolyn Hogg and Katherine Belov

The science of de-extinction has gained attention over the past decade, driven by technological advances that bring the likelihood of resurrecting extinct species ever closer. De-extinction is achieved via one of three approaches: back-breeding, cloning or genetic engineering (Shapiro 2017). With an extinct species such as the thylacine, a reference genome can be used to answer questions surrounding biology, evolution and the possibility of de-extinction (Shapiro 2017; Feigin *et al.* 2018). Essential to this is an accurate and complete map of protein-coding genes within the genome, termed 'gene annotation'.

The thylacine genome was published in 2018 and sequenced from a 108-year-old preserved pouch young (Feigin *et al.* 2018; see pp. 42–4). Because of the degraded nature of the ancient DNA, the thylacine genome is fragmented and contains many ambiguous bases or gaps where the DNA sequence could not be assembled. This makes identifying the location of genes within the genome challenging, especially for gene families with a complex genomic organisation and high level of variation, such as those involved in immunity (Horton *et al.* 2004).

Knowledge of thylacine immunity is currently limited to histological and anatomical descriptions of preserved pouch young (Old 2015) and our understanding of immunity at the genetic level is extremely limited. Compatibility between thylacine clone and surrogate dam (likely to be Tasmanian devils) at functionally important immune genes is one of many components that must be considered for de-extinction. In eutherian mammals (i.e. humans, domestic animals), immune receptors called major histocompatibility complex (MHC) and uterine natural killer (NK) cells play an important role in maternal recognition and the immune response to pregnancy. Although marsupial placentation differs to that of eutherian mammals, maternal recognition of pregnancy and immunity remain important components of reproduction in this mammalian lineage (Griffith *et al.* 2019).

Immune genes are some of the most variable regions of the genome, because they need to generate high levels of diversity to recognise and respond to an extensive range of pathogens (Horton *et al.* 2004). Given this, characterising the genomic location and organisation of immune genes is challenging, even in genomes containing full-length chromosome sequences such as the current human assembly. Characterising thylacine immune genes and their similarity to those in Tasmanian devils is essential for determining the feasibility of de-extinction in the future.

The thylacine genome contains all major immune genes encoding receptors (T cell receptors (TCR), MHC, uterine NK receptors and toll-like receptors (TLRs)), signalling molecules (cytokines) and antibodies (immunoglobulins (IG)) required for a functioning immune system (Peel *et al.* 2021). These include genes involved in innate immunity, such as TLRs that recognise patterns on the surface of pathogens, thereby activating an immune response. We also identified genes involved in adaptive immunity, such as immunoglobulins and TCRs (Peel *et al.* 2021).

Immunoglobulins, commonly known as antibodies, bind to foreign antigens leading to pathogen destruction. T cell receptors expressed on the surface of T lymphocytes bind and present antigens to other immune cells, leading to activation of an immune response. Marsupials contain a number of unique innovations within their immune system, which was also the case for the thylacine (Belov *et al.* 2007). We identified the ancient T cell receptor TCRμ and large expansions in NK receptor gene families in the thylacine (Peel *et al.* 2021), both of which are not found in eutherian mammals but are present in other marsupials such as Tasmanian devils, koalas and the gray short-tailed opossum.

At the genome level, the thylacine immune gene repetoire was very similar to the Tasmanian devil (Morris *et al.* 2015). However, at a gene level, genomic regions encoding immune genes were highly fragmented which limited our ability to determine similarity between the thylacine and Tasmanian devil at these sites. Variable regions within MHC, IG and TCR that are involved in recognition of non-self, maternal-fetal compatibility and antigen binding were highly fragmented. Although partial gene sequences were identified, complete genes could not be reconstructed.

The current fragmented nature of the thylacine genome sequence does not include complex, highly variable immune gene sequences. Our understanding of functionally important immune genes is incomplete, and hence the molecular information necessary to support de-extinction strategies. At this stage, it is not possible to compare loci or variation within immune genes between dam and clone. Any future genetic engineering will need to be able to detect slight differences in genomic sequences between the thylacine and Tasmanian devil, and this will require a more contiguous thylacine genome.

Sexual dimorphism and behaviour in marsupial carnivores

Nicole Dyble

In most mammalian groups males tend to be larger than females due either to intrasexual (within sex) selection and competition for mates (e.g. males fighting among themselves for mating opportunities) or epigamic (between sex) selection in which females choose among males to obtain mating partners. Large body mass may facilitate the evolution of sexual dimorphism by making polygyny (a pattern of mating in which a male has more than one female mate) more likely and thus increasing the intensity of sexual selection between sexes. Sexual dimorphism also arises from natural selection by favouring different optimal size for males and females because they occupy different ecological niches (Ralls 1977), thereby reducing direct competition between the sexes for resources. Tasmanian devils (*Sarcophilus harrisii*) and quolls (*Dasyurus* spp.) are all sexually dimorphic.

Male devils tend to have a thicker neck and broader head and chest once they reach peak maturity because female devils typically solicit the larger, dominant males and are known to reject smaller males, both physically and vocally. The male needs to be in good condition with enough body mass to sustain him through mating over a prolonged period of time, because he has to physically force his potential mate into submission. Females are often less broad through the chest with a more pointed or triangular jaw. There is no long-term association between males and females after mating has occurred.

During their early weeks of life, females are in fact heavier than males, but by 18 months this pattern is reversed and thereafter males increase rapidly in mass (Guiler 1978). Although males can be sexually mature at 3.5 kg (Pemberton 1990), some continue to grow until they are 3–4 years old, by which time they may reach their final adult mass of 8–14 kg. In 1993 the weights of 100 animals of each sex from the north-east coast of Tasmania were reported, with males averaging 10.2 kg and females 7.1 kg (Pemberton and Renouf 1993). Tasmanian devils living in higher altitudes were smaller, with males at 8.43 kg and females 5.4 kg (Jones 1997), suggesting that thylacines may have exhibited a similar pattern.

Tasmania is also home to two of the six species of quoll. Sexual dimorphism in eastern quolls (*Dasyurus viverrinus*) and spotted-tailed quolls (*Dasyurus maculatus*) is very pronounced, with males up to 1.5 times the mass of females and with stronger canines for their body size (Jones 1995). Eastern quoll males average 1.1 kg vesus 0.7 or 0.8 kg for females (Green 1967), although weights up to 1.9 kg for males are known (Bryant 1988). The selection pressure for larger body mass in males may be maintaining a large home range and physical competition for females. Spotted-tailed quoll males have home ranges approximately three times the size of the females'. They also measure approximately 38–76 cm from head to body, with a 37–55 cm tail length and average 3.5 kg (up to 7 kg) in body weight. Females measure approximately 35–45 cm from head to body length, 34–42 cm in tail length and average 1.8 kg (up to 4 kg). They are also the only quoll to have spots on their tails.

Tasmanian devil behaviour

The Tasmanian devil is an opportunistic carnivore and ambush predator. They have a distinctive 'lope', which they can use over long distances in order to conserve energy while wearing down their prey. They will ambush prey, injure it and wait until the weak or injured animal wears itself out while being chased. Alternatively, with smaller prey, such as birds, they just catch them and eat them on the spot (observed in captivity while birds have been perched on water bowls, for example). The exception to this seems to be the Tasmanian forest raven (*Corvus tasmanicus*), which will often place itself in very close proximity to the carcass and the devils, waiting for them to leave in order to feast on the leftovers. Perhaps captive devils, being fed regularly, are not motivated enough to attack and eat these large corvids.

Watching devils lope can be deceptive, as it does not show their true speed. They can travel very quickly at a full gallop, especially when chased away from a carcass by another devil. In captivity, when new additions are eventually introduced into an enclosure, they have been observed being chased or followed by another devil, both animals adopting 'cruise control' in order to be economical. Devils lose heat through their ears, which are full of capillaries (making them appear redder when they have an elevated heart rate) and nose, rather than panting or sweating, so they can be prone to heat stress.

Devils use sophisticated physical and chemical signals, as well as an array of vocalisations to establish themselves around a carcass. Females tend to be dominant for most of the year (the exception is soliciting and submitting to males during the breeding season). She will often have joeys to feed somewhere in a den, or will be carrying pouch young, so needs to provide milk and meat for them.

Devils housed together in captivity over a period of time have been seen to form bonds with other devils. They will often 'groom' each other, particularly the females. Some females have been observed greeting each other by licking around the jaws and face. Females have also been seen to groom males, with him lying down while she licks his neck and facial region. The animals that are very familiar with each other don't tend to vocalise or posture as much while approaching food. They are not challenging different wild devils every day and tend to settle down to feeding with limited, if at all, vocalising and posturing. These animals will also often den together (Plate 6). Wildlife parks often feed their devils during the day, leading to more diurnal activity.

Devils can also form bonds with their keepers. Other interactions include very confident devils bluffing and chasing their keepers, sometimes while their backs are turned. Often, when confronted, the devil will retreat. Other times, a shoe may be fastened onto and gnawed.

Devils can consume up to 40% of their body weight in a sitting. Captive devils have more controlled feeding cycles. Gorging such a large amount of meat in a short period often results in the animal having a distended stomach and laying down to digest it. A devil can be easier to approach and be potentially vulnerable in this state. It is likely that the absence of other large predators has resulted in feeding this way, even when thylacines were Australia's largest carnivore. This may support the belief that thylacines may have eaten only soft tissue, leaving the rest to the devils.

Devils have a less narrow food base. They have a squat, muscular body and short strong legs enabling loping for long hours in search of food and reproductive partners (especially males). The devil has an abundance of whisker beds acting as sensors during movement, feeding and communication. Devils have no milk teeth. Instead, their teeth erupt at approximately 4 months of age and remain for the devil's lifetime, becoming quite worn in old age. Their claws and feet are designed to dig and climb efficiently. Sense of smell is acute and they can detect food up to 1 km away. The body structure of a devil (particularly older males) includes a shark-like forward torso, with a large broad head and neck, especially after attaining their peak maturity (Plate 7).

Conflict over a carcass tends to be avoided through ritualised behaviour. Feeding devils communicate with each other through a range of visual postures, vocalisations and various chemical signals. Juveniles are often crepuscular; the older, more confident devils often active overnight. Their relative, the spotted-tailed quoll, however, is an active diurnal feeder. Male and female devils have similar-sized home ranges, which is unusual in sexually dimorphic, solitary carnivores. Large males obtain the additional food they need by eating for longer (Owen and Pemberton 2005).

PART 2: EVOLUTION, PALAEONTOLOGY AND TAXONOMY

Origin and early evolutionary history of marsupials and their relatives

Russell K. Engelman

Although marsupials today are regarded as a primarily Australian lineage, they have a long history on other continents and one-third of all living marsupial species are actually in South America. Remains of fossil marsupials and their relatives are known from all seven continents, with some dating back as far as the Cretaceous (145.5–66 Mya). The oldest known proto-marsupials come from the early Cretaceous (120 Mya) of North America (Cifelli and Davis 2015). However, given that stem-placentals have a fossil record dating back to the late Jurassic (~160 Mya), stem-marsupials must have diverged from placentals before this time but remain undiscovered.

Marsupials underwent a significant radiation on the northern continents, in some cases, such as in North America, even exceeding placentals in species diversity and morphological disparity (Wilson 2013). However, most of these lineages died out in the Cretaceous–Paleogene extinction (66 Mya), leaving only two highly generalised families in the Cenozoic: the Herpetotheriidae and Peradectidae (as well as the unusual Anatoliadelphidae in what is now Asia Minor, which was an island landmass at that time). Marsupials remained minor components of northern Cenozoic faunas for several million years before dying out approximately 11 Mya, in the middle Miocene (Furió *et al.* 2012). Several species of living opossums are known from Mexico and Central America, with one, the Virginia opossum (*Didelphis virginiana*), ranging as far north as Canada, but this is a relatively recent (<1 Mya) re-dispersal into North America from South America and the expansion of the Virginia opossum north of the Gulf Coast may have even happened within the last 100 000 years (Tyndale-Biscoe 2005).

Proto-marsupials dispersed from North America to South America in the latest Cretaceous or earliest Paleocene (75–65 Mya). No Mesozoic marsupials from South America are known, but Paleocene (66–56 Ma) South American fossil sites have produced a diverse array of marsupial species. Marsupials in South America diversified into a number of forms, including small kangaroo-like bounding 'seed-eaters' and sabre-toothed predators that may have weighed as much as 100 kg (Eldridge *et al.* 2019). Much of this diversity has since gone extinct or is currently restricted to relict distributions in small areas of the Andes Mountains, with only the opossums (order Didelphimorphia) maintaining sizeable diversity in the lowlands. Originally this decline was attributed to competition with placentals, but it appears as though cooling climates throughout the Cenozoic are a more likely culprit (Prevosti *et al.* 2013; Goin *et al.* 2016; Engelman *et al.* 2017).

From South America, marsupials spread southward into Antarctica and eventually Australia, aided by these three continents being physically linked and Antarctica not having extensive glaciers until ~35 Ma. There appears to have been a selective barrier preventing unrestricted migration into Australia, given that many fossil mammal groups known from Antarctica and South America (including several groups of marsupials, litoptern and astrapotherian ungulates and possibly even dryolestoids; Gelfo *et al.* 2019) are unknown from Australia. Similarly, Australia must have had a diverse non-marsupial, non-placental mammal fauna before the arrival of marsupials in the early Cenozoic, but little is known about them. The Australian Mesozoic mammal record is limited to monotremes and mammals currently considered to be monotreme relatives (*Ausktribosphenos, Bishops, Teinolophos*) plus the multituberculate *Corriebataar marywaltersae.*

The early evolution of marsupials in Australia is still poorly understood. There is only one terrestrial mammal-producing fossil site in Australia that predates the late Oligocene: the early Eocene (~55 Ma) Murgon site in Queensland (Beck 2013). Six species of mammals have been described from Murgon: *Djarthia murgonensis*, an early relative of modern Australian marsupials; *Thylacotinga bartholomaii* and *Chulpasia jimthorselli*, two species that have been closely linked to marsupials in South America; *Archaeonothos godthelpi*, a small faunivorous marsupial of uncertain affinities; at least one marsupial more closely related to South American marsupials as represented by an anklebone; and *Tingamarra porterorum*, a possible 'condylarth' placental.

Notably, none of the Murgon marsupials can be assigned to any extant Australian marsupial order. The first representatives of Dasyuromorphia (thylacines, dasyurids and numbats), Diprotodontia (kangaroos and allies) and Peramelemorphia (bandicoots and bilbies) do not appear until the late Oligocene (<26 Ma), though genetic evidence suggests they have been distinct since the Eocene (56–33.9 Mya). In the case of Dasyuromorphia, thylacinids (thylacines), dasyurids (quolls and allies) and numbats (myrmecobiids) do not appear to have diverged from one another until the Eocene–Oligocene boundary (~34.7 Mya) and both early thylacinids and the other major group of large marsupial carnivores in Australia, the thylacoleonids (marsupial lions), appear to have been relatively small (up to 25 kg). This raises the question of which animals filled the ecological role of large terrestrial carnivore in Australian ecosystems prior to the late Oligocene.

It is unclear if these niches were filled by reptiles (e.g. mekosuchine crocodiles), especially as goannas (the primary terrestrial competitors of carnivorous marsupials today) do not appear to have been present in Australia until the late Oligocene (~27 Mya; Vidal *et al.* 2012), or if there was a now-extinct group of large carnivorous Australian marsupials that have yet to be discovered. Such a discovery would not be unprecedented. The Murgon *Archaeonothos* does not appear to be a dasyuromorphian (Beck 2013) and Kealy and Beck (2017) found that the putative Oligocene dasyurids *Ankotarinja tirarensis* and *Keeuna woodburnei* may form a clade only distantly related to Dasyuromorphia, suggesting there may have been more groups of Paleogene faunivorous marsupials in Australia that, similar to the thylacine, have since become extinct.

Evolutionary relationships of Australia's carnivorous marsupials (order Dasyuromorphia)

Shimona Kealy and Robin Beck

The marsupial order Dasyuromorphia comprises over 80 predominantly carnivorous and insectivorous species living in Australia and New Guinea today (Flannery 1995; Van Dyck and Strahan 2008). This species richness makes Dasyuromorphia the second most diverse order of Australian marsupials after Diprotodontia (kangaroos, possums and their relatives). Modern dasyuromorphians include quolls or 'native cats' (*Dasyurus* spp.), the numbat or 'marsupial anteater' (*Myrmecobius fasciatus*), antechinuses (*Antechinus* spp.), dunnarts (*Sminthopsis* spp.) and a small hopping carnivore of Australia's desert regions, the kultarr (*Antechinomys laniger*). They also include the world's smallest living marsupial, the long-tailed planigale (*Planigale ingrami*, body mass ~4 g) and the largest living carnivorous marsupial, the Tasmanian devil (*Sarcophilus harrisii*, body mass >8 kg). Another member of Dasyuromorphia was the thylacine (*Thylacinus cynocephalus*), which was, until its recent extinction, the largest carnivorous marsupial to survive into historical times (weighing 15–20 kg; Rovinsky *et al.* 2020). Australia's fossil record reveals, however, that there were once multiple species and genera of thylacines that spanned a broad range of sizes (see pp. 34–6).

Although the smaller dasyurids share some superficial similarities with the opossums (family Didelphidae) of the Americas, and the thylacine has in the past been compared with the South American sparassodonts (an extinct group of carnivorous marsupial relatives), detailed anatomical and genetic analyses show that Dasyuromorphia is more closely related to other Australian marsupials than to those of South America (Horovitz and Sánchez-Villagra 2003; Duchêne *et al.* 2018). In particular, DNA evidence indicates that dasyuromorphians are most closely related to bandicoots and bilbies (order Peramelemorphia), and the enigmatic marsupial moles (order Notoryctemorphia), which are specialised burrowing marsupials that today live in the deserts of Central Australia (Duchêne *et al.* 2018). This grouping of dasyuromorphians, bandicoots and bilbies, and the marsupial moles has been named Agreodontia, which comes from the ancient Greek words for 'hunter' and 'tooth', referring to the fact that most of its members are carnivorous or insectivorous.

Within Dasyuromorphia, three modern families are currently recognised (Jackson and Groves 2015): Myrmecobiidae (which includes a single species, the numbat *Myrmecobius fasciatus*), Thylacinidae (which includes a single modern species, the thylacine *Thylacinus cynocephalus*, but also several extinct species) and Dasyuridae (all the remaining modern species, i.e. devils, quolls, antechinuses, dunnarts etc.). A fourth, entirely extinct family is known, Malleodectidae, which includes two fossil species with enormous crushing premolars that might have fed on snails and other hard-shelled prey items (Archer *et al.* 2016).

Of the three modern families, Thylacinidae is thought to be the first group to diverge from the common ancestor of Dasyuromorphia, perhaps ~30–35 million years ago (Kealy and Beck 2017). During the late Oligocene and early Miocene, thylacines appear to have thrived in the wet, rainforest-like environments that characterised Australia at this time (Kealy and Beck 2017; Rovinsky *et al.* 2019). However, ~14 million years ago there was a major drop in temperature around the world (known as the middle Miocene Climatic Transition), which caused Australia to become much drier, reducing the extent of rainforest habitats and replacing them with more open environments: firstly woodlands and later, grasslands and deserts. This major change in climate coincides with a significant decline in thylacine diversity and also a substantial radiation of dasyurids, which remain highly diverse and successful today (Kealy and Beck 2017; Rovinsky *et al.* 2019).

This change in evolutionary fortunes may have been due to climate change causing the extinction of most thylacinids and freeing up ecological niches for the dasyurids to evolve into. Alternatively, the dasyurids may have been better adapted to the new, open habitats, enabling them to thrive and directly outcompete the previously dominant thylacines. Whichever scenario is correct, by ~5 million years ago only a single genus of thylacine (*Thylacinus*) remained, while dasyurids had diversified into multiple different groups (Kealy and Beck 2017; Rovinsky *et al.* 2019). The further opening up of habitats in Australia over the past 5–6 million years, and the development of Australia's first grasslands, coincided with the rise of the smallest dasyurids, the planigales (Kealy and Beck 2017).

By 3 million years ago the modern thylacine (*Thylacinus cynocephalus*) was the last surviving member of a distinctive and once much more diverse family of carnivorous marsupials (Rovinsky *et al.* 2019). Meanwhile, members of the family Dasyuridae had radiated to become Australia's dominant marsupial carnivores, filling virtually every available ecological niche, from sandy deserts, to cracking clays, to open woodlands and tropical rainforests.

The evolutionary story of Australia's carnivorous marsupials tells of the diversity and abundance of thylacines that once ruled the ancient rainforests, of major climactic change that saw the decline in dominance of the thylacines and the ascendance of a new carnivorous family, the dasyurids, and of the strong links between climate change and the evolution and extinction of Australia's marsupials. Finally, this story emphasises the historical vulnerability of the last surviving species of thylacine, *Thylacinus cynocephalus*, and the tragedy that was its final, human-caused extinction.

The search for the thylacine's beginnings: fossil relatives and evolutionary history

Douglass S. Rovinsky

Big things have small beginnings. The thylacine was by far the largest marsupial carnivore in the world at the time of its extinction in the early 20th century and many of its characteristics – size, dog-like body and predatory ecology – may seem completely unique and out of place among its modern marsupial relatives. Like its relative the numbat, though, the thylacine was the last member of a long evolutionary lineage. Unlike the numbat, however, there is a record of much of that lineage, offering a glimpse into the evolutionary history of the thylacine. And for much of that history, the members of that group were small.

Thylacinidae – the family composed of *Thylacinus* and its closest relatives – split from the lineage leading to the numbat and dasyurids (~33 Mya during the late Eocene–early Oligocene; Kealy and Beck 2017; see pp. 33–4). This is a rather substantial split; about the same amount of time separates the thylacine and the Tasmanian devil or numbat as does the African lion and meerkat (Zhou *et al.* 2017). At least 12 species in eight genera are currently recognised within the family Thylacinidae, the majority of which have been recovered from early to mid-Miocene age deposits in the Riversleigh World Heritage Area in Queensland (23.0–11.6 Mya; Rovinsky *et al.* 2019). Older undiagnostic specimens have been recovered from the late Oligocene Pwerte Marnte Marnte locality, Northern Territory (Murray and Megirian 2006), with most of the younger material recovered from the late Miocene–early Pliocene Alcoota and Ongeva Local Faunas at Alcoota Station, Northern Territory (Rovinsky *et al.* 2019). The youngest non-*Thylacinus cynocephalus* specimens – mostly fragmentary and undiagnostic – derive from New Guinean and southern–eastern Australian Pliocene-aged sites.

Through the first 20 million years or so of their evolutionary history, during the Oligocene through the middle Miocene, the thylacinids were Tasmanian devil-sized or smaller, with most weighing somewhere between 2 and 7 kg (Rovinsky *et al.* 2019). By the late Miocene, however, these smaller thylacinids were almost completely replaced by much larger species. The late Miocene and younger members of the genus *Thylacinus*, four of the five known thylacinids from this time period, weighed well over 10 kg. The only smaller-bodied thylacinid from this later period, the 'broken dog' *Tyarrpecinus rothi*, is known entirely from a single, fragmentary specimen from late Miocene Alcoota Station deposits (Murray and Megirian 2000).

The genus *Thylacinus* itself is initially represented by the small-bodied (~8 kg), early Miocene *Thylacinus macknessi* from the Riversleigh World Heritage Area (Muirhead 1992). By the late Miocene–early Pliocene, arguably the most famous of the fossil thylacinids appear: the very large *T. megiriani* and *T. potens*. These two were very robust – both larger than the modern thylacine, with both species commonly estimated at >40 kg (Wroe 2001; Yates 2014). However, the fragmentary nature of the fossils, coupled with the issues of extrapolating body size across species with different proportions/builds, may be leading to an overestimation of these species' body mass (Yates 2014; Rovinsky *et al.* 2020). The latest occurring non-modern *Thylacinus* is the relatively small (<18 kg; D. S. Rovinsky, unpublished) *T. yorkellus*, from Pliocene deposits in South Australia (Pledge 1992; Yates 2015). Though only known from a partial mandible and isolated molar, it can still be seen that *T. yorkellus* was very similar to the modern thylacine, though a little smaller and with slight differences in craniodental proportions.

Early, smaller thylacinids were probably rather ecologically similar to modern quolls. The smaller of the species, such as *Muribacinus gadiyuli* and *Badjcinus turnbulli*, were likely unspecialised predators of invertebrates and small vertebrates. Larger-bodied early thylacinids, like the 4–8 kg *Nimbacinus dicksoni* and *Ngamalacinus timmulvaneyi*, probably took a greater percentage of vertebrate prey, with *Nimbacinus* having a robust skull capable of withstanding the stress of tackling larger prey (Attard *et al.* 2014). Even so, these thylacinids still had a utilitarian set of dental characteristics, with teeth able to slice flesh, crush insects, grind plants and chew bone all relatively well, but specialising in none of those things. There was, however, another lineage of thylacinids.

The genus *Thylacinus*, together with the closely related, early Miocene *Wabulacinus ridei*, break from this 'one-size-fits all' style of craniodental adaptation. Even at the earliest stage of the lineage, these thylacinids show several craniodental characteristics derived towards carnivory (vertebrate eating), such as reduction of tooth complexity and longer, straighter shearing crest (Rovinsky *et al.* 2019). These dental characteristics, which show an increase in cutting efficiency over grinding, as well as a dramatic increase in body size, suggest that this lineage was operating in a different functional niche than the other thylacinids. More and more vertebrate flesh was being eaten; thylacinids were becoming specialised vertebrate hunters – hypercarnivores. This trend reached its maximum derived state in the lineage with the modern thylacine (*Thylacinus cynocephalus*).

The thylacine was the last surviving tip of a diverse branch extending back over 30 million years. Many of its seemingly unique features can be traced back through this lineage and indeed, several traits are shared in kind with its closest relatives. We are lucky to have this record, unlike, for example, the numbat – another modern singleton among the Australian marsupials. When the thylacine was lost, an entire family and all its history was snuffed out; hopefully, the numbat will not share the same fate.

Thylacine footprints in the fossil record

Aaron Camens

Separating thylacine footprints from those made by other animals such as dogs and foxes, or even Tasmanian devils and *Thylacoleo carnifex* (the marsupial 'lion'), is a big challenge. Although the pad morphology of the thylacine foot has some significant differences to that of dogs, it is very rare that fossil footprints preserve the key details that allow us to distinguish exactly which species we're looking at. Fossil footprints are thus often just vague oval outlines and the characteristic spacing between consecutive prints is used to identify the track maker.

When a newly discovered fossil footprint is evaluated the first thing to consider is the age of the sediments housing the footprint – are the rocks 5000 years old, 500 000 years old or 5 000 000 years old? It is the age of the footprint that is most important when we want to know the animal that made the tracks. If the rocks are 5 000 000 years old, that is before the thylacine (*Thylacinus cynocephalus*) even evolved, although there was an even larger extinct thylacinid, *Thylacinus potens*, alive at that time. If we fast forward to 5000 years ago, then things get tricky because it is thought that dingoes arrived in Australia somewhere around this time and thus footprints from this period could have been made by a thylacine or a dingo.

Palaeontologists who study fossil footprints and other traces preserved in the fossil record are known as 'ichnologists'. In Australia there are very few people working on fossil footprints from the time period when thylacines walked the Earth. Fortunately for those of us who do, nearly all of the known fossil footprint sites predate the arrival of the dingo, meaning that it is relatively easy to narrow down the track maker if something has left dog-like prints. The only site that is potentially problematic is Nyereena Jinna, near Clare Bay in South Australia. This amazing place preserves the footprints of humans, emus, wombats, kangaroos and, most likely, dingoes but the age is likely somewhere in that critical period when thylacines and dingoes overlapped in mainland Australia.

Most of the fossil footprints that have been identified as belonging to thylacines occur in the fossil sand dunes of the Bridgewater Formation and the Tamala Limestone along the coast of southern Australia. The northern-most occurrence is near Shark Bay in Western Australia, but the best-preserved prints come from coastal outcrops of the Tamala Limestone and from Jewel Cave, both in south-west Western Australia. Additional prints from Kangaroo Island in South Australia (Camens *et al.* 2017) and Warrnambool in south-west Victoria are also known.

In most cases the tracks were imprinted at a time when the sea level was at or near where it is today, which means that they are likely from the last interglacial period, somewhere ~110 000–140 000 years ago. Associated with these tracks a whole range of other animal tracks have been found, including animals still extant today, such as emus, wombats, quolls and kangaroos, and extinct megafaunal species such as the giant marsupial *Diprotodon optatum* that weighed 2 tonnes and the extinct, flightless bird *Genyornis newtoni* that stood over 2 m tall.

These fossil trackways are important for documenting when and where thylacines lived in the past, but they contain other important information, such as how they moved. Did they run or mostly walk? Did they trot or bound like dogs or did they have other gaits unique to marsupials? Were they solitary or did they move around in groups? Through the study of these prints we can learn things about the thylacine that we can't learn anywhere else.

The search for the scant record of *Thylacinus* in north-west Australia

Cassia Piper, Peter Veth and Carly Monks

The thylacine has remained an enigmatic part of Australian wildlife, often described as shy, quiet and only fleetingly seen. In death, the archaeological and palaeontological record of this animal is scarcely more visible, yet tantalising fragments of bone and teeth can be found in cave deposits, in both the zooarchaeological and palaeontological context, raising hope that more information can be gleaned about this elusive creature.

Zooarchaeology is a multidisciplinary branch of study that analyses the remains of animals found in archaeological sites and often overlaps with palaeontology to provide records and knowledge on fauna. In this chapter, we look at several archaeological sites from north-west Australia that have contributed to the scant record of the thylacine.

The Montebello Islands are an archipelago of islands 70–90 km off the north-west coast of Australia. The archipelago's position at the edge of the continental shelf means that during periods of Pleistocene low sea level, particularly during the Last Glacial Maximum 25 000–19 000 years ago, the coast was within easy reach of the land that makes up today's Montebello Islands. Over two field seasons in 1991 and 1993, Peter Veth and colleagues sought to investigate the archaeological potential of the archipelago, particularly in regard to the possible preservation of Pleistocene marine resource records, which are generally thought to have been submerged in many coastal regions by rising post-glacial sea levels (Veth 1993; Veth *et al.* 2007). A group of three caves was found on a peninsula on the eastern side of Campbell Island. Two caves, named Noala and Hayne's Caves, contained archaeological material and were analysed in the 1990s. There was no archaeological material found in the material from Morgan's Cave, so it was stored in the Western Australian Museum palaeontology collection (Veth *et al.* 2007).

The remains from Morgan's Cave were reanalysed in 2013 and 2014. The most abundant species present were rodents and small marsupials (Piper and Veth 2021), but also within the assemblage were two small fragments of thylacine teeth, indicating the species was at least present in the area. The fragments were identified using well-preserved specimens of the thylacine in the Western Australian Museum collection, enabling comparison of the structure of the teeth (Piper and Veth 2021). To understand why only two broken teeth were found, taphonomy was used to investigate how the remains ended up in the cave. This included investigating the accumulating agents responsible for collecting animal bones within the cave, such as owls or humans, by looking at the type of animals found in the deposit and how well preserved they were. For example, barn owls regurgitate pellets, resulting in generally whole skulls and jaws, as well as other bones, of small to medium-sized animals (Andrews 1990). Owls from the family Strigidae, such as the southern boobook (*Ninox boobook*), don't do this, so the bone remains from their prey are fragmented and broken (Worthy and Holdaway 1996).

Unfortunately, even with the help of taphonomy, not much can be gleaned from the thylacine teeth found in Morgan's Cave. They were near the bottom of the deposit, which suggests at least 3000 years old, and it is likely that the top layers were accumulated after that, because the thylacine was by then extinct on the mainland at ~3200 years before present (White *et al.* 2018b). Additionally, these small islands are unlikely to have supported mid-sized marsupials. The find likely predates 7000 years ago when the island was connected to the mainland (Veth *et al.* 2007). The high number of rodent remains indicates it was likely that owls, not thylacines, were the main predator contributing to the deposit. The teeth themselves are relatively unworn and only consist of tooth caps with no tooth roots, meaning they were from young animals (A. Baynes, *pers. comm.*). Whether the young were there with parents, or as prey, we cannot say.

The Montebello Islands, Barrow Island and Cape Range are all part of the same geological region, once connected in a long stretch of coast and hinterland. The thylacine has been recorded at several sites within Cape Range, including Mandu Mandu Creek Rockshelter and Monajee Cave (Baynes and Jones 1993) and more recently at another rockshelter close to Mandu Mandu Creek Rockshelter, as part of the ongoing Nyinggulu Archaeology Project. The most recent find includes several fragments of thylacine bone, teeth and jaw apparently belonging to a single individual, possibly one that died naturally within the cave rather than being brought to the site by a predator or through misadventure.

Based on the first radiocarbon dates returned from the site, the thylacine teeth and skeletal elements probably date to the mid-Holocene, similar to those from Morgan's Cave. Here, however, they are clearly associated with human occupation that spanned most of the Holocene and up to as recently as 205 ± 18 BP. As at Morgan's Cave, the thylacine remains from the Cape Range sites are poorly preserved and the bones and teeth are very fragile. Where the archaeological record shows that the Montebello Islands were abandoned by people *c.*7000 BP, Cape Range shows repeated human occupation from before 40 000 years ago up until European contact.

Unfortunately, the thylacine is currently elusive and enigmatic in north-west Australia. The remains from zooarchaeological, palaeontological and rock art sites show it was at least present in the area, but how it was behaving and interacting with other animals, and even humans, remains a mystery for now.

Thylacine from Nombe and Kiowa Rock shelters, Papua New Guinea

Mary-Jane Mountain

Nombe is a small rock shelter on the slopes of Mt Elimbari in the Eastern Highlands of Papua New Guinea near the patrol post of Chuave. It was excavated first briefly in 1964 by Peter White under the name of Niobe and then for several seasons between 1971 and 1980 by Mountain. Both excavations were part of doctoral research from the Department of Prehistory, Research School of Pacific School, Australian National University (White 1972; Mountain 1991). Mountain divided the sediments from Nombe into broad strata: stratum A at the surface to stratum D at >2 m below the 1980 surface in some parts of the site. Thylacine remains were recovered from all four strata. Table 1 summarises the main deposition history of the site with bracketing dates for the four strata (Denham and Mountain 2016).

The thylacine (*Thylacinus cynocephalus*) represents the largest indigenous mammalian predator known to have existed in the highlands within the period of human occupation of New Guinea. The largest number of thylacine faunal specimens was recovered from the lowest levels at the site in stratum D. This was acknowledged in a 1983 publication of the extinct species recovered from stratum D (Flannery *et al.* 1983) and more recently another specimen was identified by experts from Flinders University (Van Zoelen *et al.* 2020). There was evidence for single animals in strata A, B and C. There is slight variation in the sizes of the animals, which are usually smaller than many Australian specimens (L. Dawson, *pers. comm.*). The most recent thylacine bone was recovered from stratum A (Sutton *et al.* 2009), as was a dog mandible, which indicates that the introduction of the dog and the extinction

Table 1. Broad sediment deposition model for Nombe (based on table 1 in Denham and Mountain 2016)

Stratum D is subdivided based on occurrence relative to a former stream channel.

Stratum	Bracketing dates (cal. BP)*	Sediment deposition and human activity
A	Present–5650	Loose and friable, ashy sediments containing much broken animal bone and stone artefacts
B	6300–10 400	Dark brown loams with many artefacts and heavily burnt broken bone
C	13 500–16 400	Red-brown to brown sediments with flowstones and tephra blocks with sparse artefacts
D (1/5)	19 600–25 500	Red-brown clays deposited in a previous stream channel containing animal bone and some stone artefacts
D (2–4)	29 900–38 700	Basal red-brown and ginger clays, which are sediments in the banks of the previous stream channel and contain some animal bone and possible human artefacts

* Radiocarbon dates have been corrected for changing levels of atmospheric carbon over time, into uniform calendar years where BP (Before Present) is relative to 1950AD.

of the thylacine both occurred within the last 5650 years. More detailed dating has not been possible so far.

Other extinct species were recovered from stratum D. Flannery *et al.* (1983) designated these as *Protemnodon nombe*, *Protemnodon tumbuna* and *Dendrolagus noibano*, as well as the remains of an unidentified diprotodontid. Further work since then (Kerr and Prideaux 2022) has identified further remains of *Protemndon tumbuna*, identified a new species (*Nombe nombe*), changed the identification of the *Dendrolagus noibano* remains to more likely represent *Dendrolagus dorianus* and identified some of the unknown diprotondid remains as being *Hulitherium tomasettii*. There were human artefacts recovered from the same stratum, but no reliable association or clear evidence to identify whether humans had hunted the animals or they died by natural means.

A thylacine scapula from the base of stratum D2 has been dated by direct uranium–thorium) dating to a minimum of ~126 ka and the *Hulitherium* to ~55 ka (Prideaux *et al.* in press). A thylacine mandible was also recovered by Sue Bulmer from her excavations in 1959 at a neighbouring rock shelter site of Kiowa (Bulmer 1966; Gaffney *et al.* 2021). This was identified by Van Deusen (1963). The thylacine bone came from Zone 2A, Level 9, which broadly dated to the early Holocene. Some of the best-preserved thylacine specimens from Nombe are now in the Australian Museum, Sydney. Other specimens are now part of the Mary-Jane Mountain Collection in the School of Archaeology and Anthropology, College of Arts and Social Studies, Australian National University.

The presence of thylacines at both Nombe and Kiowa in Holocene occupation levels alongside much evidence of human hunting from broken and burnt animal remains, as well as frequent flakes and other stone artefacts, indicates they were occasionally hunted and caught by communities in the Holocene period. No such clear and unambiguous evidence was recovered from the Pleistocene levels at Nombe. Indeed, it is possible that the site was used by thylacines at some stages as a lair and that some of the other animal bones, possibly including specimens from extinct species, were the remains of thylacine prey. A formidable animal indeed!

Pups of the Swan Coastal Plain

Kailah M. Thorn

The last place we would expect to find Australia's youngest mainland Tasmanian 'tiger' is south-western Australia, the opposite side of the country from its name-state. Yet some of the youngest palaeontological specimens of the thylacine (*Thylacinus cynocephalus*) were collected from cave deposits north of the Swan River. A stretch of Pleistocene calcarenite, known locally as the Tamala Limestone, is the equivalent of the Bridgewater Formation that surrounds much of coastal southern Australia. A karst (cave) system throughout the Tamala Limestone runs parallel with the coast from Wanneroo (inclusive of the Yanchep Caves) within the Perth Metropolitan Area and continues north through Drover's Cave National Park near Jurien Bay for more than 200 km. These relatively young caves have sparsely sampled Western Australian late Quaternary fauna and preserve a record of the larger extinct marsupial carnivores.

Inspired by the remarkable megafauna finds early in the 20th century from Mammoth Cave near Margaret River, the caves of the Swan Coastal Plain were palaeontologically prospected by a young Fulbright scholar (a foreign exchange scholarship program between Australia and the USA) from Texas, Dr Ernest Lundelius. Many of the caves he explored on the Swan Coastal Plain contained remains of locally extinct macropodids (kangaroos and their relatives) and the coprolites of Tasmanian devils. The first reports of these fossils hinted at the possibility that there could be other, larger predator remains in these caves.

Lundelius' curiosity was followed up soon after by Western Australia's active speleological society, the Western Australian Speleological Group (Inc.), with their members reporting a number of fossil finds to the WA Museum (WAM). A jaw and other bones belonging to at least two individual thylacines were found in Mystery Cave (J-6) in the Jurien area, by keen spelunker Lindsay Hatcher in 1986. Harry Butler, namesake for the WAM's Harry Butler Research Centre, discovered the bones of a thylacine in a cave near Stockyard Cave (E-3) in the Eneabba karst region. In the South Hill caves area, Wedges Cave – more likely a denning site of the Tasmanian devil rather than *Thylacinus* – produced a single molar tooth as evidence of the larger predator. All these fragments are in geologically young karst formations, with the fossils unlikely to have accumulated before the Late Pleistocene (126–11.7 ka). Relatively young fossils in comparison with those found elsewhere on mainland Australia.

The first large-scale excavation of a cave on the Swan Coastal Plain was during Dr Alex Baynes' postgraduate research in the 1970s, on Hastings Cave in the northern-most stretch of the Plain near Jurien Bay. This site preserves a late Pleistocene–Holocene account of faunal change in the area before the arrival of the dingo, but notably does not contain any *Thylacinus* remains.

A single right third metatarsal (WAM 13.11.364) of a thylacine was uncovered from Caladenia Cave between Gingin and Guilderton (Plate 8), roughly 1 hour north of Perth. With an associated charcoal age of 3254–2925 years, this specimen could be the youngest record of mainland thylacines in Australia. But these fossils aren't just pups in the sense of their relatively recent deaths. Archaeological excavations in the Yellabiddie Caves just north of Jurien Bay recovered elements of multiple juvenile individuals and an adult molar, collectively suggesting a den site from >5000 years ago. Caladenia Cave also produced a juvenile near-edentulous (toothless) dentary (Plate 9).

A common feature of each of the *Thylacinus* finds in the cave deposits on the Swan Coastal Plain is the presence of multiple young individuals. Many of the teeth and edentulous (toothless) jaws are those of joeys or young adults, more than likely sheltering in the cave. The Tamala Limestone karst once hosted a population of growing thylacine families. These creatures were living within cooee of Perth on Noongar boodja land only 3000 years ago.

The thylacine genome and the genetic basis of adaptive evolution

Charles Y. Feigin

The image of the Tasmanian tiger (thylacine) has captivated people for millennia, with representations ranging from ancient Aboriginal rock paintings to the mascot of the Tasmanian men's state cricket team. Its innumerable depictions reveal a strange carnivore with a dog's head, tiger's stripes and a pouch like a kangaroo. Perhaps the most gripping are the small number of haunting, black-and-white film strips documenting some of the last individuals of this enigmatic species, during their final years in captivity. The very existence of such films is a reminder that even in 2022, the tragic extinction of the thylacine is still within living memory.

The thylacine's recent extinction and the abundance of preserved museum specimens around the world (see pp. 1–2) has long tantalised biologists with the possibility of analysing its DNA. During my doctoral studies, I had the immense privilege of taking up this task. But beyond personal fascination with the thylacine, what broad benefit to science is there in studying such a peculiar animal? As it turned out, the project began during a significant debate in evolutionary biology that the thylacine was uniquely suited to contribute to.

With the completion of the Human Genome Project (HGP) in 2001, biology entered the genomics era. Techniques developed during this project rapidly made it feasible for individual laboratories to analyse the whole genomes of wild species. Genetics was no longer restricted to traditional model organisms such as mice, zebrafish and fruitflies and it became possible to test decades of accumulated evolutionary hypotheses by directly examining the signatures in DNA.

Well before the HGP, biologists had determined that genes (short stretches of DNA encoding the instructions for building proteins) were vital components of an organism's genome. Proteins build tissue, digest food, facilitate metabolism, manage cell division and act as the molecular 'hardware' that reads the DNA code. Thus, it was widely assumed that adaptive evolution proceeded primarily through changes in protein-coding genes. However, genome comparisons across diverse animals revealed that many genes seemed to change very little, even over millions of years. Paradoxically, this property (called sequence conservation) is due in part to the critical roles of protein-coding genes. Indeed, their functions are so important that mutations altering their structure or function are often deleterious. Moreover, many genes are recycled, playing roles in the development of multiple traits (known as pleiotropy), so even a random mutation that improves one trait may harm another. Thus, the evolution of protein-coding genes is often highly constrained (Carroll 2008).

However, it was also found that protein-coding genes comprise only a small fraction of most animal genomes (in humans, only 1–2%). Some considered these large, non-coding regions to be 'junk DNA', though the pervasiveness of this belief is often overstated. By the 1980s it had already been shown that some non-coding regions contained functional DNA called cis-regulatory elements (CREs) that act as control modules for nearby genes (Gillies *et al.* 1983). Many CREs only act in one or a few places in an organism's body, suggesting that they may be less pleiotropic (and therefore less constrained) than the protein-coding genes they regulate (Carroll 2008).

During the extended thylacine genome project, we sought to address the relative contributions of protein-coding genes and CREs to adaptive morphological evolution. Two key factors made the thylacine an exquisite model species for this purpose. First, it is a remarkable example of convergent evolution, having a skull nearly identical to that of placental canids (e.g. wolves; see pp. 7–9). Because the thylacine and canids are very distantly related, they represent two independent, natural experiments in producing a single skull morphology. Second, the skull is a shared ancestral trait in all mammals that was present in their last common ancestor. Because of this, many of the genes and CREs involved in skull formation are also shared among living mammals. This allowed us to directly compare the 'same' protein-coding gene and CRE sequences from the thylacine, canids and many other mammal genomes and identify signatures of natural selection at the DNA level. Our findings were quite startling.

Although we identified clear evidence of natural selection acting on both thylacine and canid protein-coding genes, we found that the specific genes differed in both species and included few skull-related genes (Feigin *et al.* 2018). In contrast, we found much more widespread evidence of natural selection in CREs, including many shared between the convergent carnivores (Feigin *et al.* 2019). Our findings were one part of a large number of studies revealing similar findings. Today, evolutionary biologists are digging ever deeper into the evolution of CREs in diverse species. Importantly, the field has adjusted its assumptions in the light of such comparative genomic studies and shifted focus to understanding the different contexts in which various classes of DNA elements contribute to adaptation.

The thylacine genome project is one of many recent examples of how studying non-model species can clarify fundamental principles of biology. However, DNA cannot teach us everything. In addition to robbing the world of a beautiful and ecologically important species, the unforgivable extinction of the thylacine means that many exciting questions will go unanswered. Worryingly, some of the thylacine's closest relatives, such as the Tasmanian devil, numbat and eastern quoll, are themselves on the brink. For those unable to see the intrinsic value in the weird and wonderful species that still exist, my great hope is that recognition of their immense scientific value will be justification for their protection.

Examining the thylacine's first extinction using ancient DNA

Lauren C. White

The thylacine is an infamous example of human-driven extinction. Photographs and films of thylacines from the first few decades of the 20th century are a poignant reminder of what we have lost, and yearly through National Threatened Species Day on 7 September we commemorate the day the last known thylacine died at Beaumaris Zoo. However, this was not the first extinction experienced by the thylacine. Although in contemporary history thylacines were only found on the island of Tasmania, they were once also widespread on the Australian mainland. The fossil record tells us that mainland thylacines became extinct ~3200 years ago, at the same time as mainland devils (White *et al.* 2018b; see pp. 45–7).

The cause (or causes) of these first extinctions are much debated. The three hypotheses that are most frequently put forward are based on changes in Australia's environment around the time of the extinction and Tasmania's isolation from their likely effects. Firstly, the dingo – a potential competitor and predator of thylacines – arrived in Australia ~5000 years ago, but never reached Tasmania as rising sea levels had flooded Bass Strait thousands of years earlier (Corbett 1995). Secondly, the archaeological record shows that human population size and resource use expanded during the Holocene, but this trend was absent in Tasmania (Johnson and Wroe 2003). Finally, during this same period, the Australian climate shifted towards a drier and more drought-prone system, a change that is assumed to have been less severe in Tasmania due to its maritime environment and more consistent rainfall (Donders *et al.* 2008).

Unravelling a species' ancient history is a difficult business and revealing the cause of the thylacine's mainland extinction has been, and continues to be, tackled using many lines of evidence. I and my colleagues at the Australian Centre for Ancient DNA have used DNA extracted from ancient and historical thylacine specimens to contribute to solving this mystery and increasing our understanding of these enigmatic animals.

DNA extracted from old bones, teeth and skin can be challenging to work with because it is typically fragmented, damaged, present in very low quantities and contaminated by the DNA from bacteria and other extraneous sources (Dabney *et al.* 2013). The thylacine samples I worked with came from both historical specimens collected during the bounty years and much older specimens, from animals that died up to 20 000 years ago. These ancient samples were collected from cave sites across southern Australia, where cold and dry conditions allowed DNA preservation.

Because the DNA in these types of samples is present in such low quantities, it is essential not to further contaminate it, so we work in ultra-clean laboratories wearing full protective suits and decontaminate all equipment regularly (Llamas *et al.* 2017). We further increase our chances of retrieving comparable sequences across samples by targeting the thylacine's mitochondrial genome, the part of each animal's genome that is inherited from its mother and is present in a higher number of copies per cell than nuclear DNA (Stoneking 2018).

Using the 51 mitochondrial genomes that we sequenced from mainland and Tasmanian thylacines we could infer a number of interesting insights about thylacine history (White *et al.* 2018a). Mainland thylacines had separated into two populations, in the east and west, before their extinction and the long-term population size of the western group was larger and more genetically diverse than the historical Tasmanian population. Our results also suggested that, prior to the flooding of Bass Strait, thylacines traversed the land bridge between eastern mainland Australia and Tasmania.

In agreement with other genetic studies (Menzies *et al.* 2012; Feigin *et al.* 2018), our results showed that Tasmanian thylacines had low genetic diversity immediately prior to their recent extinction and that a number of lineages (haplotypes) had been lost well before European arrival. This indicates that the Tasmanian population contracted sometime in the past. Using phylogenetic analyses, we estimated the timing of this bottleneck to be ~3000 years ago, roughly coincident with the mainland extinction. Interestingly, a similar pattern has been shown in Tasmanian devils (Brüniche-Olsen *et al.* 2018).

We think this temporal congruence of the Tasmanian and mainland decline is significant and tells us something about the thylacine and devil mainland extinctions. Because dingoes never reached Tasmania, and there is no evidence of changed resource use by Indigenous Tasmanians in this time period, we are left with climate change as a potential driver of the prehistoric Tasmanian bottleneck. This firstly means that the effect of climate change on mainland thylacines and devils may have been more severe than previously thought, as this climatic shift is assumed to have been less acute in Tasmania. Secondly, the contrasting outcomes of these declines (i.e. mainland extinction and Tasmanian survival) suggest that the mainland thylacines and devils were under additional stress, implicating the existing pressures of the dingo invasion and increased human resource use in their extinction.

Thus, we propose that all three hypothesised causes – dingoes, humans and climate change – played a role in the thylacine's mainland extinction. These results are very pertinent to biodiversity conservation today, because almost all of Australia's native animals are now under immense and increasing pressure from invasive species, human resource use and climate change. We should take heed of the thylacines' ancient and recent history and work to avoid more extinctions of our unique wildlife in the future.

Diagnosing a synchronous extinction

Lauren C. White, Frédérik Saltré, Corey J. A. Bradshaw and Jeremy J. Austin

Both thylacines and Tasmanian devils were once broadly distributed across mainland Australia, as evidenced by cave art in Kakadu National Park, fossilised footprints on Kangaroo Island

and fossils found from the west to the east of the continent (Brandl 1972; Camens *et al.* 2017; Peters *et al.* 2019). By European settlement in 1788 both species were extinct on the mainland, with the remaining populations only found on the island of Tasmania until 1936 when the last known thylacine died in the Beaumaris Zoo. Based on the youngest-dated fossils available, both mainland thylacines and devils were often assumed to have gone extinct on the mainland at roughly the same time (as well as the Tasmanian native hen, *Tribonyx mortierii*), ~3500 years ago (mid-Holocene; Johnson and Wroe 2003). This assumption – that their mainland extinctions occurred synchronously – has led to the hypothesis that these population declines had a common cause.

Suggested drivers of these two extinctions in mainland Australia are competition with the dingo (a potential predator and competitor), changes in human subsistence practices and/or climate change. These three suggested causes are based on estimated timelines of major events and the isolation of Tasmania from their likely effects. For example, the dingo arrived in Australia ~5000 years ago (i.e. ~1500 years before the assumed date of the mainland thylacine extinction), but could not cross Bass Strait unassisted to reach Tasmania (Prowse *et al.* 2014), potentially protecting the insular population of thylacines from being outcompeted for prey species by the dingo (see pp. 20–3). But the direct killing of mainland thylacines by dingoes also likely occurred (Letnic *et al.* 2012).

The second hypothesis is based on changes in human population size and subsistence practices on mainland Australia during the Holocene (Johnson and Wroe 2003). These changes, revealed through the archaeological record in many regions of mainland Australia, are referred to as 'intensification' because they indicate more intensive use of natural resources by humans (Lourandos 1997). In the mid- to late-Holocene a wide variety of more sophisticated stone tools appeared, including hunting weapons (Lourandos 1997). Concurrently, the human population of mainland Australia grew and expanded into previously unoccupied habitats (Johnson and Wroe 2003). These changes could have affected the thylacine and devil through direct hunting and increased encounters with humans or by a reduced abundance of their prey species. In Tasmania, however, there is no evidence of late-Holocene advances in technology or unprecedented human population growth (Lourandos 1997).

The mainland thylacine and devil extinctions also coincide approximately with a shift in Australia's climate, suggesting a third possible driver: following a climatically stable period, a strengthening of the El Niño–Southern Oscillation (ENSO) cycle caused drier, more drought-prone weather in Australia beginning ~6000 years ago and intensifying by ~3000 years ago (Donders *et al.* 2007). Rainfall subsequently became more annually variable and instead of seasonal or yearly lean periods, thylacines and devils might have had to contend with multi-year events of low resource availability (Brown 2006). Many extant Australian mammals survive such 'bust' periods at low population densities and in refuges of favourable habitat, but this strategy is less suitable for large predators such as thylacines and devils that typically have large ranges necessary to acquire prey (Letnic *et al.* 2011; Prowse *et al.* 2014). Tasmania is likely to have been largely spared the effects of ENSO-related climate change due to its maritime climate ensuring more consistent rainfall (Donders *et al.* 2007).

These three hypotheses – dingo arrival, human intensification, and climate change – are not necessarily mutually exclusive, but disentangling their relative importance to the extinction of mainland thylacines and devils is a challenging and ongoing area of research. One of the main tools used to investigate these extinctions are the dated fossils themselves, because they reliably indicate where and when a species was present; the youngest-dated fossils available therefore reveal the last time a species was 'observed'. With this information we can build timelines of events and infer the most likely causes of ancient extinctions. However, the youngest fossil age of an extinct species is almost never a good proxy for a species' true extinction date. Fossils are extremely rare because the alliance of specific conditions needed to preserve an organism from scavengers, digesting microbes and erosion are uncommon. It is therefore improbable that the last living thylacine or devil on the mainland died under circumstances that would have allowed its body to fossilise and then be found by scientists (Signor and Lipps 1982). Additionally, not all published fossil ages are equally reliable and all are inherently uncertain, with the margin of error increasing the farther back in time that the organism died.

With all these challenges, how can we be sure that mainland thylacines and devils truly went extinct synchronously and how can we robustly assess the cause of their extinctions? Fortunately, new scientific developments over the past 10 years have provided the opportunity to tackle these challenges. The *FosSahul* database is the first to compile the ages of non-human vertebrate fossils for the past 180 000 years in the Sahul region (the combined landmass of Australia and New Guinea joined during periods of low sea level). This database comes with a quality-rating algorithm enabling the exclusion of unreliable fossil ages (Peters *et al.* 2019). Many statistical methods have also been developed to estimate the time gap between the last dated fossil and the final extinction date to account for uncertainty while building extinction chronologies (Saltré *et al.* 2015).

The latest updated reliable fossil dates (based on 56 devil specimens and 48 thylacines) and statistical analysis reveal that the mainland devil and thylacine extinctions occurred at almost the same time, 3200 years ago (±50 years) (White *et al.* 2018b). This extinction 'window' is similar to what had been previously assumed, so it does not challenge or support any particular hypothesis about the cause of the declines. However, by supporting the predominate assumption with reliable data and rigorous statistical methods, this analysis provides a strong and defendable foundation on which future work can build.

Continuing research into past extinctions is increasingly important in this time of unprecedented biodiversity decline. If we can uncover the factors that led to extinctions in the past, we will be better placed to combat the current crisis of species loss today.

Genetic diversity in the Tasmanian tiger

Brandon R. Menzies

General scientific intrigue with the Tasmanian tiger combined with the government bounty scheme introduced from 1888 (which resulted in thousands of animals being killed) led to a large amount of thylacine material being scattered around the world in various institutions and now available for genetic (and other scientific) studies. There are at least 812 individual thylacine specimens across 23 countries, with most of these samples outside Australia, demonstrating how sought after these specimens were (see pp. 1–2). Although many of these samples can be used to extract DNA for analysis, much of this material is of very poor quality due to its age, degradation, method of preservation etc.

One scientific question that is relatively easy to decipher using thylacine biological material such as skin, muscle or hair and subsequent DNA is how genetically diverse the species was prior to extinction. Being genetically diverse as a species is important because it gives the species a larger repository of potentially unique genes with which to fight threats such as new diseases. For example, one reason it is thought that the Tasmanian devil suffers from the deadly facial tumour is its limited genetic diversity within the population generally (Morris *et al.* 2013).

To assess genetic diversity, we can look at a number of different genes. Most commonly, researchers use mitochondrial DNA (mtDNA) diversity, which is particularly helpful in studies of museum-archived specimens because it is more abundant generally than nuclear DNA and the specimens generally have poor DNA profiles due to age and degradation. In 2012, my colleagues and I published a paper describing the genetic diversity of mtDNA from a study of 12 museum-archived thylacine specimens collected between 102 and 159 years ago (Menzies *et al.* 2012). To do this we sequenced this specific DNA and counted the differences between specimens in the four DNA bases: As (adenine), Cs (cytosine), Ts (thymine) and Gs (guanine).

To highlight the considerable problems of working with museum-archived specimens we actually sampled 22 specimens in total but could only acquire usable DNA from 12 of these (or 54.5%). Furthermore, the region of mtDNA that is normally amplified in studies such as ours to compare genetic diversity is riddled with what are known as repeat units in the thylacine (see pp. 170–2). This made the comparing of fragmented pieces of DNA much more difficult.

What we discovered from our analyses was that the thylacines that existed in Tasmania were genetically very similar to each other prior to their extinction. To be precise, they were 99.5% similar between individuals. By comparison this is ~25% of the genetic diversity seen between individual wild dogs and wolves (Vilà *et al.* 1997). Put another way, our diversity estimates for the thylacine were similar to some of the least diverse populations of the endangered Leadbeater's possum, another marsupial (Hansen *et al.* 2009).

We don't (yet) understand why the thylacine had such low genetic diversity, but it may be due to reductions in the size of the population prior to the isolation of Tasmania from mainland Australia, as has been shown to exist now for the Tasmanian devil (Morris *et al.* 2013). Although we can't predict how this low genetic diversity might have affected the Tasmanian tiger since European settlement, due to its extinction, it may have resulted in similar threats that are now facing the Tasmanian devil, such as transmissible disease.

A brief taxonomic history of the thylacine

Branden Holmes

Today it is widely accepted that only a single species of the genus *Thylacinus* survived into the Late Pleistocene (*c.*126–11.7 ka), the last member of its entire taxonomic family. Both the diversity of recent thylacines, and their distinctiveness among other carnivorous mammals, have been the source of much discovery, disagreement and debate over the past 200 years. The species was originally given the scientific name *Didelphis cynocephala* (dog-headed opossum), alongside the Tasmanian devil (*Didelphis ursina*) or bear opossum, by George Prideaux Harris (Harris 1808), the Deputy Surveyor of Van Diemen's Land. Thus, they were both placed in the genus *Didelphis*, with the American opossums, in the family Didelphidae.

It was soon recognised by the eminent French embryologist, palaeontologist and comparative anatomist, Étienne Geoffroy Saint-Hilaire, that the Australasian members of the genus constituted a separate lineage. Consequently, he moved them to a new genus, *Dasyurus* (Saint-Hilaire 1810). The differences between the two genera became apparent to the German biologist and palaeontologist Georg August Goldfuss, who created the family Dasyuridae for the Australasian species (Goldfuss 1820). The thylacine was subsequently recognised by Coenraad Jacob Temminck, the Dutch ornithologist and zoologist, as being distinct enough to warrant its own genus, *Thylacinus,* although he used the junior synonym *Thylacinus harrisii* in his description (Temminck 1824–27).

The final major taxonomic change was the recognition by the Italian biologist Charles Lucien Bonaparte that the thylacine was unique even at the family level. He consequently created the family Thylacinidae for the species in 1832 (Bonaparte 1832) and not, as I have discovered, 1838, which has been inaccurately accepted until now (S. Jackson, *pers. comm.*). Other scientific names for a living thylacine were published, such as *Thylacynus striatus* (Burnett 1830), but there is no evidence that these authors actually thought that a second living species existed.

An expedition led by Thomas Mitchell in 1830 to explore the Wellington Caves in New South Wales collected some of the country's first known fossils, which included possible fossils of the thylacine (*The Sydney Morning Herald* 1842). Indeed, the famed Victorian-era anatomist Sir Richard Owen described the first of three fossil species now regarded as being synonymous with the thylacine, the Cave thylacine (*Thylacinus spelæus*), from 'one of the caves in the Wellington Valley' (Owen 1845). The earliest described fossil species of thylacine that is still considered valid, the Powerful thylacine (*Thylacinus potens*), was only published in the late 1960s (Woodburne 1967).

In 1868, the Director of the Australian Museum, Gerard Krefft, published the description of a second living species of thylacine, the 'Bulldog-Tiger' (*Thylacinus breviceps*) (Krefft 1868). He claimed that older Tasmanians themselves distinguished between the two species, with the more familiar species being the 'Greyhound-Tiger'. The Tasmanian naturalist and solicitor Morton Allport questioned the distinctiveness of the new species, but added that he too was familiar with the name Short-nosed or Bulldog tiger (Allport 1868). An exhaustive search of earlier published accounts fails to corroborate either author's claim, which may have been in unpublished sources such as private letters. The German thylacine expert Heinz Moeller noted that the specimen Krefft credited as being a separate species was actually an immature specimen of the modern species (Moeller 1968). Consequently, the references to 'bulldog' or 'greyhound' tigers are most likely local terms for the male and female of the species.

During the 1970s and 1980s, several scientists studied fossils of the modern thylacine from mainland Australia to test the hypothesis that these remains represent more than one species/subspecies. They each concluded that the individual variation in the material does not warrant the scientific description of a new thylacine that died out in the relatively recent past. However, a much more recent treatment of the hypothesis has indicated that in fact there may be more than one kind of thylacine represented (Helgen and Veatch 2015), which would further compound the recent loss of the family Thylacinidae. Yet, it is necessary to look afresh at a subject when new methods and evidence accrue, no matter how uncomfortable the conclusion, because only then will we have a real chance at making our future choices better than our past mistakes.

PART 3: ABORIGINAL KNOWLEDGE AND ARCHAEOLOGY

The relevance of rock art in understanding the thylacine's mainland extinction chronology

Ken Mulvaney

Thylacine images in the rock art of Aboriginal Australia are common right across the continent, recognised by archaeologists long before it was generally understood that this animal once roamed throughout mainland Australia and was not limited to Tasmania (Figs 5–7). The introduction of the dingo is implicated in the demise of the thylacine, a proposition supported by the rock art. Dingo images are all associated with recent rock art less than 4000 years old, whereas thylacine depictions link to earlier artistic traditions.

It is across northern Australia where the majority of thylacine depictions are found, partly because of the abundance of rock art, correlating with stable ancient rock surfaces and preservation of the art. In some places, such as the Dampier Archipelago on the north-west coast of Australia, such images can be traced through many tens of thousands of years of rock art production.

Given the problems associated with dating rock art, the presence of extinct fauna has been a useful, although somewhat controversial, temporal indicator. One of the problems is that for most of the extinct fauna there are only partial skeletal remains and paleontological reconstructions as a guide to their appearance. With the thylacine, there are photographic

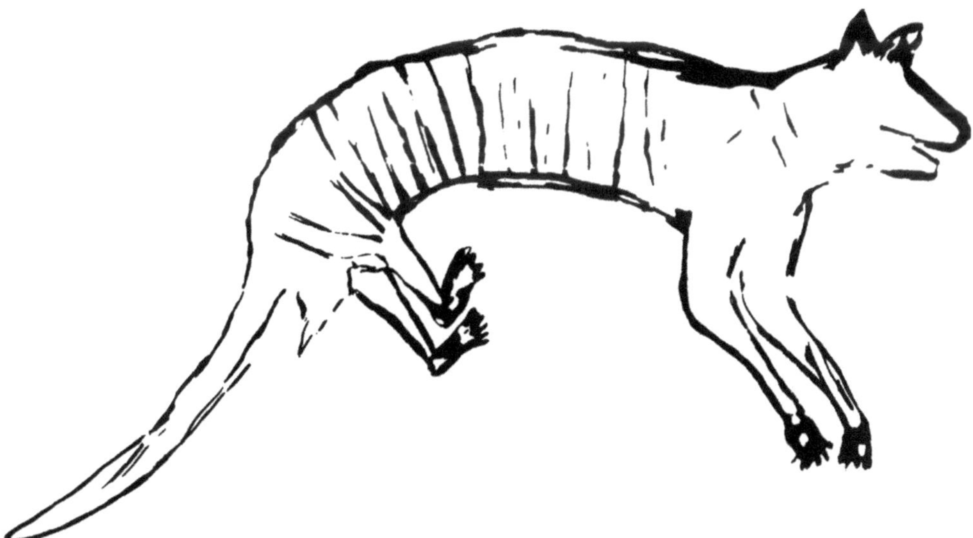

Fig. 5. Painted Arnhem Land example.

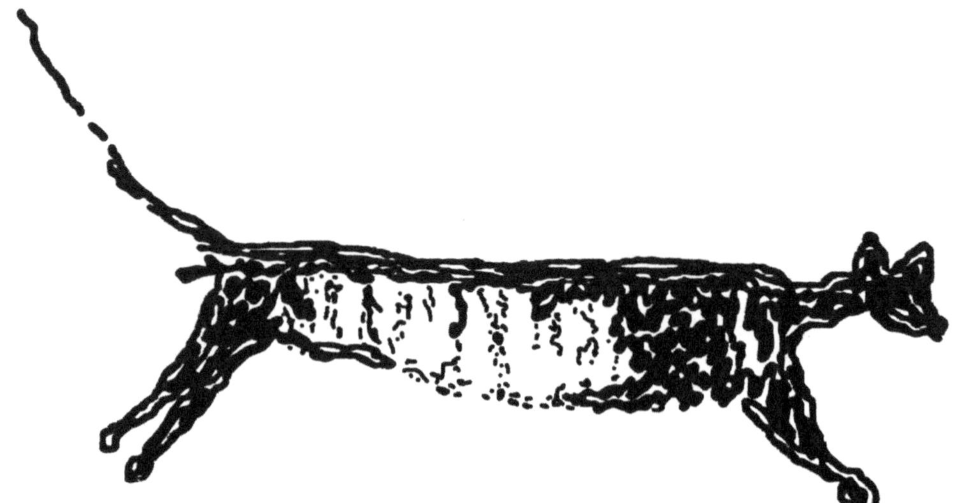

Fig. 6. Painted Kimberley example.

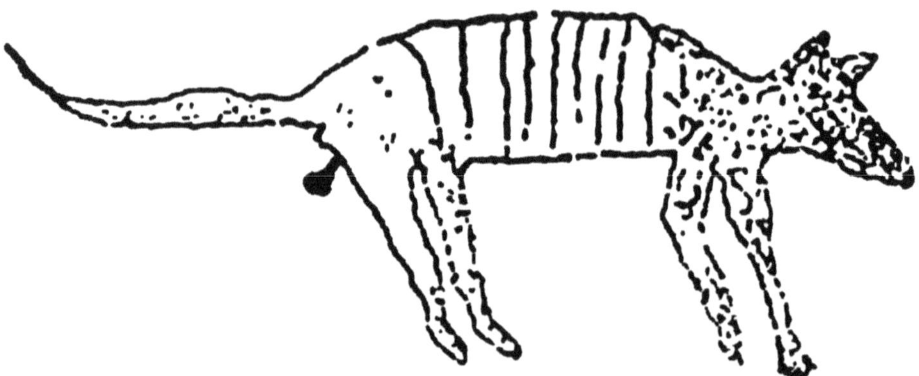

Fig. 7. Engraved Pilbara example.

images and museum specimens to inform the interpretation of the rock art images. It is clear that the artists were familiar with the animal, rendering the image in naturalistic form and anatomical detail.

In both engraved and painted art, from ancient times up until they became extinct, the distinctive form of the thylacine is identifiable within the many styles and artistic phases of Aboriginal Australia. The diagnostic anatomical character of this animal, in addition to the patterned stripes, includes the elongated body, relatively massive head, short, rounded ears, hind- and forelimb proportions and the long, tapering tail, all features that in combination are not present in other Australian fauna. An additional attribute is the tuft of hairs at the tip of the thylacine tail, which has been emphasised as a significant indicative trait in the finely executed paintings of the Kimberley and Arnhem Land regions.

Both the artistic conventions of style and technique, together with the physical properties of the rock surface influence how the rock art image appears. Painting with a fine brush allows for detail such as the tuft of hairs and enables flexibility in forming the shape of ears. For engraving (petroglyphs), these subtle details are much more difficult to produce, especially on extremely hard rock surfaces. The absence of explicit anatomical detail is usually linked to the specific graphic dictates of the culture and not necessarily related to specific traits of the animal being represented. Thus the absence of body stripes may not mean absence of thylacine, especially when the other diagnostic attributes are present.

In the case of the Dampier Archipelago, Pilbara region, where some 50 petroglyphs of the thylacine on the distinctive basaltic block formations are known, the panel on which the images are placed imposes physical restrictions. In these cases, it appears that the size and shape of the rock panel influenced the particular design solutions, such as exact limb length ratio and the angle and length of the tail. It is also possible to observe the changing artistic phases and associated graphic conventions within thylacine depictions that span from well before the last Ice Age to when the animal became extinct, some 30 000 to just 3000 years ago.

Within the rock art images there is not just a record of the animal, there is associated meaning and story linked to the image. In one example, with the tail up rather than straight-out or slightly lower than the animal's back is to accord with the depiction of this animal urinating or marking territory (Fig. 8). A mating pair is depicted on a rock in the inland Pilbara and young are depicted together with adult animals. There are examples, such as in Arnhem Land, where human figures are included with the animal and these appear to record Aboriginal people hunting the thylacine. The painting of the tail tuft may be a signalling device used by the artist, possibly mimicking thylacine behaviour.

Debate remains as to exactly when the thylacine became extinct – before 3000–3500 years ago on mainland Australia and 1936 in Tasmania – with some people believing it still survives.

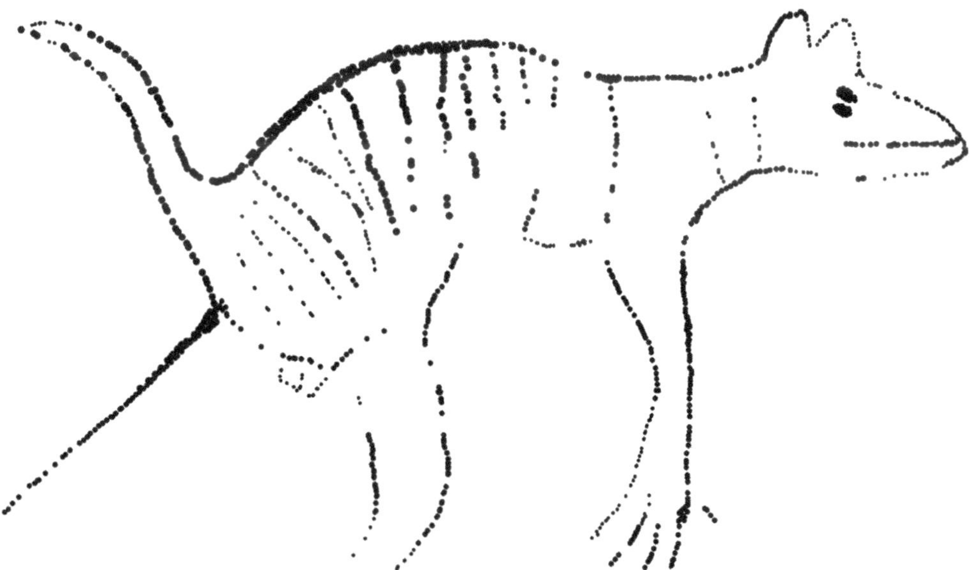

Fig. 8. Pecked and abraded Angel Island case with tail up.

Fig. 9. Charcoal drawn example from Wollemi.

Its presence in the rock art certainly points to the earlier demise; however, a charcoal drawing of what certainly looks like a thylacine was found within a rockshelter in the Wollemi area of the Blue Mountains in New South Wales (Fig. 9). Although the age of this drawing is unknown, the rock art technique is not thought to have been used in the general region prior to 1600 years ago (Taçon *et al.* 2011). If this is correct, it is possible that either the animal did survive here longer than in other parts of mainland Australia or archaeologists are wrong about when the particular rock art technique was used.

Extinction, inscription and Dreamings: some mainland thylacine connections

Katie Glaskin

Thylacines are estimated to have become extinct on the mainland 3000–3500 years ago (White *et al.* 2018b). Consequently, gaining information about the nature of Aboriginal relations with this long extinct marsupial on the mainland is difficult. This necessarily brief review should be read in conjunction with other available overviews (Glaskin 2021; Knights and Langley 2021), as well as emerging overviews such as the forthcoming documentary *Tiger on the Rocks* (Fiske and Vasseleu 2022). It is also likely that more Aboriginal narratives about the thylacine will emerge in the future for reasons detailed more fully in Glaskin (2021).

Australian Aboriginal cosmologies share the concept of a creative period in which ancestral beings shaped the country (now widely known to English-speakers in and beyond Australia as 'the Dreaming'), with associated narratives describing their activities. The ancestral beings are thus intrinsically connected with named places and topographic features; a common theme of many narratives is transformation, from human to animal to topographic feature. Partly for this reason, some rock art depictions may be regarded by local Aboriginal people as having been made by ancestral beings; some may also be regarded as an instantiation of the being depicted and as evidence of its ongoing material presence (Taylor 1996). Given this, many of the rock art depictions from around mainland Australia of thylacines may have, or have had, specific significance in Aboriginal cosmology.

In Western Arnhem Land, Chaloupka and Murray (1986) recorded that local Kunwinjku people identified rock art depictions of the thylacine, which appear in a naturalistic style, as djankerrk, 'dog associates of the Rainbow Snake', the latter a highly significant being in regional cosmology. Kunwinjku painter John Mawurndjul similarly described a thylacine at Milmilngkan as the Rainbow Serpent's pet (in Garde 1997). In the short community film *Thylacine & Red Kangaroo* (Grose and Hohnen 2014), also related to Western Arnhem Land, the thylacine is referred to as Mawolbbolya. One recorded narrative talks about the djanggerrk (thylacine) swimming (Chaloupka 1993), which accords with Tasmanian descriptions of thylacine behaviour (Milligan 1853).

The number of sources recording Kunwinjku narratives about the thylacine (e.g. Garde 1997; Grose and Hohnen 2014; Evans 2018; Nabarlambarl *et al.* 2018) are more extensive than have, at this point, been found for other parts of mainland Australia. The existence of rock art provides an important record and basis for maintaining an enduring connection with the animal, which although physically absent, is regarded as ontologically present, at least with respect to certain of its rock art depictions (Mawurndjul in Garde 1997). At Murujuga (the Burrup Peninsula in the Pilbara region of Western Australia), where at least 50 thylacine rock art sites have been recorded (Mulvaney 2019), it has not been possible to locate similar Aboriginal accounts concerning these depictions and their relationship to local Aboriginal cosmologies in the available ethnographic record. This is likely connected to the effects of colonisation on local Yaburara people and their population (Glaskin 2021).

In Central Australia, there is a 'wild dog story' associated with the Alice Springs area, which tells of two dogs fighting and creating named places in the area. The local dog, Akngwelye, lived in the area and the fight occurred when its territory was encroached by an unknown intruding dog from the south (Doris Stuart, cited in Stevens 1988; Finnane 2011). Since at least 2011, some local Arrernte custodians have come to describe one of the two dogs involved in the fight as a thylacine, although this 'perception' is not fully shared by other Arrernte people and contradicts previously recorded accounts (Finnane 2011; Kimber 2011; Stuart 2019). There is evidence from other ethnography (e.g. of the Flinders Ranges region) that suggests the English term 'dog' is in some cases used more generally to refer to a carnivorous animal (Tunbridge 1991) and also, in cases where a specific animal has become extinct, that it can, over time, become narratively replaced by another conceptually similar animal (Tunbridge 1991).

Given the apparent correlation between dingo arrival and mainland thylacine extinction (Johnson and Wroe 2003), there is the possibility that the thylacine may have come to be replaced by the dingo in some Aboriginal Dreaming stories concerning the former over time, in much the same way as the thylacine may now, in some cases, be 'read back' into existing narratives featuring the dingo. Notwithstanding this, there are, of course, those who would argue that the thylacine's mainland extinction date may be more recent than currently available archaeological evidence suggests.

Archaeological evidence for the Indigenous use of thylacine-based materials

Tessa Knights and Michelle C. Langley

The thylacine in the archaeological record

Despite living alongside Aboriginal Australians for at least 65 000 years, the thylacine was almost non-existent in the archaeological record until the recent discovery of a section of jawbone smeared with red ochre at the Northern Territory site of Madjedbebe (Clarkson *et al.* 2015). This jawbone is the first and only example of a thylacine part being utilised to make something by Aboriginal Australians and therefore raises the question: Why have we not found any other examples of Aboriginal Australians making things out of thylacine materials (such as bones, teeth, or skins), when the animal itself is prominently displayed in rock art, as well as paintings showing thylacines being hunted? Although it is possible that the Madjedbebe artefact is a one-off experimentation with a piece of handy material, it is far more likely we are simply yet to find other examples of worked thylacine materials in the archaeological record.

How much opportunity for interaction was there?

We now know that humans were present on the Australian continent by at least 65 000 years ago (Clarkson *et al.* 2017) and, from the fossil record, that thylacines were spread across the continent from thousands of years prior. The thylacine has been found depicted in amazing detail at multiple rock art sites in Western Australia (WA), Northern Territory (NT) and New South Wales (NSW), with at least four published images depicting thylacine hunting scenes. Interestingly, the most prolific painting of the thylacine occurs in the Arnhem Land (NT) and Pilbara (WA) regions, with both areas retaining Aboriginal stories of the thylacine into the present day. This connection, and the survival of oral history, suggests strong cultural importance of the thylacine to these communities. Unfortunately, although very persuasive, these depictions do not provide conclusive evidence that there was direct interaction between the local Indigenous groups and the thylacine, as the artists may have observed the creatures from a distance or painted fictitious events.

There is a similar issue with the thylacine remains that have been found at multiple archaeological sites around Australia. Unfortunately, these bones do not show any physical signs of being altered by humans, such as burn marks, cut marks or other intentional modifications. It is possible that thylacine were being hunted and utilised at these sites, but it is also possible that they were simply inhabiting the same area and died naturally in the area.

To work out how much opportunity there was for people to hunt or otherwise collect thylacine materials for use, we mapped all currently known human occupation sites and overlaid them with locations of dated fossil material (Knights and Langley 2021). From this you can see a large overlap of people and thylacines living in the same regions at the same time (Plate 10). Furthermore, people lived alongside thylacines for at least 37 000 years in the southern part of the continent, and with the potential for much longer in the northern part. Clearly, there was plenty of opportunity for interaction between Aboriginal Australians and thylacines, but what could people potentially use the animals for and where will we find the evidence?

How might the thylacine have been used?

With only one fragmented example of thylacine bone that was altered by humans, we can only speculate how the thylacine may have been used by Aboriginal Australians over the past 65 000 years. Looking at Indigenous use of similar-sized carnivores such as the dingo and Tasmanian devil, however, could give us an idea.

Ornamentation made from Tasmanian devil teeth is recorded from a burial at Lake Nitchie (Macintosh 1971). This necklace required a minimum of 46 animals to make, though the actual number was probably much higher. In more recent ethnographic records, ornamentation made from dingo includes dingo tail head-dresses, dingo tail necklaces and bracelets, and ornaments made from dingo hair. The Australian Museum in Sydney also contains pointed bone tools and hairpins made from dingo bone, as well as a necklace that included a dingo tooth. It may have been that prior to the introduction of the dingo ~3000 years ago and up until its extinction on the mainland, the thylacine filled a similar role to that recorded for the dingo and Tasmanian devil. This idea is supported by rock art examples that specifically picture humans hunting thylacines.

Where will we find the thylacine?

Unfortunately, for the majority of the Australian continent, the environment is hostile for the preservation of artefacts made from organic material such as bone or wood. If the soft parts of the thylacine – such as its striking pelt – were used by Aboriginal Australians, it is highly unlikely that these items would survive to be found as they would be among the first to decay and disappear. We are much more likely to find evidence of thylacine bones or teeth and then must rely on tiny traces left on them and the cultural knowledge of the community they belong to, to determine potential uses.

Currently, the amount of unidentifiable bone material found in archaeological sites is on average ~80% (Badenhorst and Plug 2011), mostly because of fragmentation removing the tell-tale markers that allow experts to distinguish between species. It seems highly likely that the evidence for the use of the thylacine is currently sitting in museum storage awaiting someone to develop a non-destructive method of determining from where the bones have come.

Aboriginal knowledge of rare and extinct mammals, including of the thylacine in the Kimberley

Andrew A. Burbidge

I first discovered the value of learning about Australia's rapidly disappearing desert mammals from Aboriginal people by reading Hedley Herbert Finlayson's classic book *The Red Centre*

(Finlayson 1935). Finlayson had discovered much about desert mammals by showing their skins to these people.

In 1974 and 1975, while working for the state nature conservation agency, after 2 years of high rainfall in the western deserts, colleagues and I did a series of rapid surveys in the Great Victoria and Gibson deserts, following this up in 1977–79 with a survey of the Great Sandy Desert. In the mid-1970s very little was known about the status of the mammals of the western deserts, but our surveys, although detecting all the small mammals we expected to find (and some we didn't expect to find), demonstrated that the medium-sized (or critical weight range) mammals, those with a body weight between ~35 g and 5500 g (Burbidge and McKenzie 1989), were either missing or we were looking in the wrong places. In 1976, with my colleague Phil Fuller, we started to find out what western desert Aboriginal people knew about the mammals of their country. In the Northern Territory, Ken Johnson was doing the same thing (Johnson and Roff 1982).

We soon learned that older people did not relate well to photographs, but immediately recognised museum puppet skins. So together with Ken Johnson and Ric Southgate from the Northern Territory conservation agency, between 1982 and 1985 we visited many communities showing mammal skins to older people, recording their local names, asking whether they still occurred in their country and finding out what was known about the species' biology. We found that almost all medium-sized mammals no longer occurred in the western deserts, that Aboriginal people saw mammals as an important and integral part of the whole environment and that they were greatly saddened by the disappearance of culturally important animals. Compiling and publishing the information readily given to us by western deserts' people has provided information about the distribution and status of Australian desert mammals that would not otherwise have been available (Burbidge *et al.* 1988).

In 1988, Phil and I were in the Kimberley as part of a biological survey of Kimberley rainforests (McKenzie *et al.* 1991). We were fortunate enough to be accompanied by Wunambal elder Geoffrey Mangolamara during visits to sites in his country. We soon discovered that Geoffrey could relate to pictures of animals, so as well as providing Wunambal names for the animals we captured or sighted; he also went through our field guides and identified other animals that occurred in Wunambal country. At the end of one session, he said we had not shown him a picture of the 'Tasmanian tiger' and he provided us with its Wunambal name – *ngaliwan* – saying that it had not occurred in his country for many generations (Mangglamarra *et al.* 1991) (the orthography used by us in 1988 differs from that used now). Later, in 1999 and 2008, linguists recording local names for animals and plants asked other Wunambal and Gaambera people about the thylacine, but no-one knew a name for it (Karadada *et al.* 2011) and unfortunately, information passed down through many generations has now disappeared as the old people have died.

Thylacines are depicted in rock art in several parts of the north Kimberley as they are in other parts of Australia (Lewis 2017). These depictions, together with the oral history presented above, leave no doubt that thylacines were once part of the mammal fauna of the Kimberley.

PART 4: EARLY EUROPEAN ENCOUNTERS (1792–1829)

Paterson's enigmatic female and other early European records of the thylacine

Branden Holmes and Chris Lee

There must have been almost daily encounters between Australia's Aboriginal peoples and thylacines all over the continent (including Tasmania) for many thousands of years. Yet their lack of written language means there are no specific, precisely dated pre-European records of the species, as rock art dating techniques necessarily include error margins due to uncertainties. The earliest reported European encounter with the thylacine is a set of footprints 'not unlike those of a tyger' found by Abel Janszoon Tasman's crew in 1642 (Smith 1981), but these are now believed to have been those of a sharp-clawed wombat (Guiler and Godard 1998; Mooney 2014). The thylacine had been extinct on the mainland for thousands of years by the time JC Craine in the de Vossenbosch reported a 'tiger' on an island off the Northern Territory in 1705 (Whitley 1970).

There are three serious candidates for direct or indirect reports of thylacines by early European maritime visitors to Tasmania prior to settlement in 1803. The first two of these both occurred in 1772 as part of an expedition lead by Marc-Joseph Marion du Fresne. The collective experiences of expedition members included having seen 'a little tiger [*qu'un petit Tigre*]' as well as 'traces of quadrupeds in different places, some of which resembled…dogs' (Paull 2011). Yet, as tantalising as these diary entries are, they lack the diagnostic traits that could put these possible records beyond dispute.

The third candidate is a record by Jacques-Julien Houtou de Labillardière on 13 May 1792, best known through the English translation referring to a ferocious white quadruped the size of a large dog and speckled with black (de Labillardière 1800), considered to be the earliest confirmed European record of the species (Paddle 2000). The reported size of the animal, comparable to a large dog, and its carnivorous morphology ('a beast of prey'), clearly rules out any other native Tasmanian animal, as it is universally accepted that the dingo never reached the island. Thus, it is a diagnostic record of the species, confirming that the thylacine was encountered by Europeans prior to settlement, despite the seemingly strange reference to the stripes as 'speckled'.

According to Paddle (2000), the key phrase '*tacheté de noir*' has been retranslated by Neil Murray to mean 'marked' or 'streaked', thus explaining the seeming discrepancy regarding the description of the thylacine's stripes as 'speckled'. As a professional historian fluent in French, Chris Lee has consulted de Labillardière's original French and come to a different conclusion regarding the most appropriate English translation:

> Original French: après nous etre enfoncé dans les bois, un quadrupède de la taille d'un gros chien sorti d'un Buisson tout près d'un de nos compagnons de

> voyage. Cet animal, du couleur blanche, tacheté de noir, avoit l'apparence d'une bête féroce
> English: after we were stuck in the woods, a quadruped of the size of a large dog came out of a bush close to one of our travelling companions. This animal, white in colour, speckled with black, had the appearance of a ferocious beast

The critical word '*tacheté*' means to mark something (adverb) or be marked (adjective), with many small marks (or dots). The adequacy of the original 1800 translation is thus sustained, posing the problem of how best to understand the seemingly flawed description of the stripes. The full context of the translation provides a plausible answer to the quandary. The observational conditions under which de Labillardière was working included both a dynamic animal ('came out of a bush'), and the likely effect of dappled light ('in the woods'), which plausibly made the animal's coat colour appear differently. As no further mention of the animal is made by de Labillardière, it is clear that he did not get a closer look, which could otherwise have righted his understandably flawed original observation.

Paddle (2000) considers a diary entry by the Reverend Robert Knopwood on 20 August 1803 to be the first reference to the species after first European settlement, but this is impossible as he was still in South Africa at the time and Risdon Cove was only settled in September 1803. The true earliest known post-settlement record of the thylacine was widely made known through a newspaper article in *The Sydney Gazette and NSW Advertiser* of 21 April 1805, which reported a male thylacine killed by dogs on 30 March earlier in the year (Paterson 1805b). Yet Lieutenant-Governor William Patterson's original letter to Sir Joseph Banks (30 March 1805), which evidently served as the basis of the newspaper article, includes physical measurements for an adult female thylacine too (Paterson 1805a). The most logical explanation is that the male killed was in the company of the enigmatic female measured, with the latter being released alive.

A diary entry (18 June 1805) by Reverend Robert Knopwood reported a group of five prisoners having seen a thylacine on the 2nd of May near Hobart, although a third record on the banks of the Huon River in the state's south from June 1805 also exists (Guiler and Godard 1998). George Prideaux Harris' original scientific description of the species in 1808 was penned on 31 August 1806; thus the animal caught in the vicinity of Hobart using kangaroo bait rounds out a quartet of the earliest known post-settlement records of the species before a hiatus. William Bullock later acquired a specimen of unknown origin in late 1810 or early 1811 for display in his London museum, which does not match any of the previously recorded specimens and thus must be distinct from them.

A further gap in records exists until 1817, with two newspaper articles each reporting the killing of a thylacine that had allegedly savaged many sheep on the respective properties close to Hobart. Two years later another thylacine was killed on a farm at Kangaroo Point (today Bellerive) on the eastern shores of Hobart's Derwent River after killing a sheep and returning to the kill. Then on 24 June 1820, it was reported in *The Hobart Town Gazette and Southern Reporter* that a stock-keeper managed to capture a female thylacine alive and presented it to Lieutenant-Governor Colonel William Sorell.

Undoubtedly, other early records of the thylacine remain to be found, while others must surely have been lost. Given the myriad of names that the species has been given, there may also be other records that are now too ambiguous to equate with the thylacine. Over time, further research into this most neglected period of the thylacine's social history will undoubtedly uncover new information that, together with other research areas, is now helping us fill in the gaps in our knowledge of the species. Only now are we truly learning what we have really lost.

William Bullock's thylacine

Stephen R. Sleightholme

The earliest known thylacine taxidermy to be displayed in a museum was that acquired by William Bullock (1773–1849), the noted 19th century traveller, naturalist and antiquarian. The specimen was an immature male and displayed at the Bullock Museum in London from 1811 until 1819, when the collection was disposed of by auction. In the 10th edition of the *Companion to the London Museum* (Bullock 1811), the thylacine is listed as the Zebra Opossum (*Didelphis cynocephala*) under the heading 'Quadrupeds':

> This animal, which is the only one known in any collection, is a native of Van Diemen's Land, where it inhabits among the caverns and rocks in the high and almost impenetrable glens of the mountainous parts of that country: it is the largest carnivorous animal yet discovered in New Holland, measuring from the nose to the end of the tail five feet three inches; it is said to be extremely voracious, which will scarcely be doubted, when it is known that the one described in the ninth volume of the Linnaean Transactions, p. 179, had in its stomach the partly digested remains of the Porcupine Ant-eater; it is said to have a short guttural cry, and appeared exceedingly inactive and stupid.

The thylacine taxidermy is absent from the 1808 *Companion* to Bullock's former Museum of Natural Curiosities in Liverpool (Bullock 1808). It also does not appear in the 8th or 9th London (Piccadilly) editions of the *Companion* published in 1810 (Bullock 1810a, 1810b). It is listed for the first time in the 10th edition published in 1811 (Bullock 1811), so it must have been procured at some point after 1810 and prior to going on display in London in 1811. As is so often the case with the provenance of these early specimens, its source remains something of a mystery, especially as it was acquired soon after Harris's scientific description of the species in 1808 (Harris 1808) and at a time when few thylacines had been encountered by the early Tasmanian settlers. It is known that Bullock frequently purchased unusual and exotic items from seafarers returning to port, so in all probability that is its likely origin.

In 1905, Graham Renshaw speculated on whether the taxidermy could possibly have been Harris's type specimen: 'It would be interesting, however, to know whether it was the same individual which in 1812 was exhibited in Bullock's Museum as a Thylacine or Zebra

Opossum, the only known specimen in any museum' (Renshaw 1905). However, Harris's type specimen was a larger male, with a total body length of 5 feet 10 inches (178 cm), whereas Bullock's specimen by comparison was smaller, at 5 feet 3 inches (160 cm). The 1812 *Companion to the London Museum* (Bullock 1812), confirms this to be so, by clearly distinguishing between the two specimens in stating: 'the one [Harris's specimen] described in the ninth volume of the Linnean transactions', confirming that the animals were not one and the same.

The Bullock collection was divided up and sold in 1819 and the thylacine taxidermy, together with the rare black emu (*Dromaius novaehollandiae ater*), was purchased by the Linnean Society of London. On 10 November 1863, the Linnean Society disposed of a significant part of its miscellaneous collections at auction, with the Curtis Museum in Hampshire (UK) being the successful bidder for the Bullock taxidermy in the sale. The taxidermy remained in the Curtis Museum's collection with certainty until 1889, when it was listed in the *General Description of Contents*, as a 'Dog headed thylacinus' on display in cases 4–6 (Moody 1889). In 2004, a handwritten copy of an annual report dated 3 December 1932 was discovered by the museum's curator, which states: 'The cases [4–6] which contained foreign material and hide, many found to be in poor condition, now display a collection of pottery' (Anonymous 1932). This statement clearly indicates that the Bullock taxidermy was no longer on display.

Today, the Curtis Museum has no thylacine in its collection, so at some point after 1889 and prior to 1932, the taxidermy was sold, exchanged or destroyed. Within the 6th revision of the International Thylacine Specimen Database, no other museum cites the Curtis Museum as the source of their taxidermy (Sleightholme and Ayliffe 2017), so regrettably, the ultimate fate of the Bullock specimen remains unresolved.

The 1819 Kangaroo Point sheep killer: natural instinct, opportunism or desperation?

Branden Holmes

Today, Bellerive is a quiet suburb on Hobart's eastern shores, but in 1819 the small settlement, then known as Kangaroo Point, was the scene of a savage attack by the Kangaroo Point sheep killer. *The Hobart Town Gazette and Southern Reporter* (1819b) reported on 24 July:

> A few days ago, an animal known in the Colony by the name of the cat tyger killed a sheep in the night on a farm at Kangaroo Point, but at the time it got clear off. Next night, determined to devour its prey, it returned to the carcase of the sheep, but was discovered by some fine kangaroo dogs belonging to the farm, and after a hard contest killed.—As the neighbourhood is very populous where this ferocious animal made its appearance, it is fortunate that no further loss has been sustained by the owners of cattle in that district; and we believe it

> is the first time, since it was inhabited, that this animal was seen so contiguous to Hobart Town.

Despite its overly feline description, the 'cat tyger' refers to a thylacine. The only other native animal that could possibly kill a sheep would be a Tasmanian devil (*Sarcophilus harrisii*), especially if the animal was weakened by disease, sickness or injury. Yet the description of a 'tyger', suggesting it may have had stripes, and the concern held for cattle, implying its large size, all but rule out the latter. It has been suggested elsewhere that this newspaper report refers to a tiger quoll (*Dasyurus maculatus*) that scavenged a feral dog kill (Paddle 2000); however, there is nothing in the report to render this remotely plausible. Moreover, as the animal put up a serious fight against kangaroo dogs, and was the first one encountered so close to Hobart Town, this confirms it to be the larger and rarer thylacine. Thus, it is a straightforward case of a thylacine killing and later attempting to scavenge, an adult sheep. To date it is the only published account of a specific instance of a thylacine killing an adult sheep prior to the first bounty on the species in 1830.[3]

In light of recent anatomical studies, even a large thylacine would have struggled to kill a lively adult sheep without risking injury (see pp. 20–3), all but ruling out natural instinct. Perhaps the sheep killed was infirm and thus a simple case of opportunism. Yet the species had not previously been recorded so close to Hobart Town, despite the many unhealthy sheep that would have provided ample opportunity for a hungry apex predator. The paucity of eyewitness accounts of thylacine predation upon livestock makes it extremely unlikely that such events were anything but rare occurrences and even then, only under extenuating circumstances.

The posthumous 2nd edition of George Barrington's book *The History of New South Wales* (1810) claimed that the thylacine 'confines its ravages to sheep and poultry'. As Barrington had died in 1804, 1 year earlier than the first post-European settlement encounter with a thylacine, an anonymous and still unknown hand evidently added the sentence for reasons unknown. Seven years later in 1817, two newspaper articles made exaggerated claims of sheep predation, which can be confidently rejected (Paddle 2000). We must then jump to the 1830s to find a published letter that speaks of four thylacines being killed 'at the sheep yard' (*The Hobart Town Courier* 1834).

There are also a number of private records of thylacines allegedly killed in the vicinity of sheep during the period (e.g. Gunn 1838), but none relate to specific instances of sheep predation and in any case they are very rare. Yet, in exonerating the thylacine from claims of being a rapacious sheep killer, it is possible to go too far in the opposite direction by seeking an alternative explanation for every single claim of thylacine predation upon livestock no matter how plausible. As the island's top predator, with a maximum body weight of ~28 kg, that at least one thylacine killed at least one sheep on a singular occasion in 1819 should not be surprising. Still, the claim that thylacines were habitual sheep killers can be as confidently rejected as the thesis that the species' extinction was natural.

---

3 The Van Diemen's Land Company's two bounty schemes ran from 1830 until their termination in 1914, with a 2-year hiatus from 1838 to 1840.

Lamb-enting a killer: the farm of Edward Abbott 'jnr'

Branden Holmes

The historical distribution of the thylacine in Tasmania included most of the island, with the putative exception of a substantial area in the south-west that was largely ecologically unsuitable for the species (see pp. 157–8). However, Sleightholme and Campbell (2016) note: 'The far south-west appears to have supported a small resident population of thylacines, primarily along its coastal fringes. Inland, the southern buttongrass plains supported few thylacines, and in all probability, this population would have been transient'.

During the 1820s, prior to the widespread killing of the species, this expansive distribution still remained. Only two major European settlements existed: Hobart in the south-east and Launceston in the north-east. It is therefore rather incredible that the epicentre of thylacine activity during the 1820s was a single farm owned by Edward Abbott 'jnr' (1801–1869) near New Norfolk in the state's south-east. Three of the six published accounts of human–thylacine interaction during the decade derive from within its boundaries north-west of Hobart.

The first event occurred in early 1821, with a very large thylacine killed on the farm, reported in *The Hobart Town Gazette and Van Diemen's Land Advertiser* for 27 January: 'A very large Hyena has lately been killed on the farm of Mr. E. Abbott, jun. at the River Plenty'. The brevity of the report is extremely interesting, as there is no mention of the animal having previously attacked any sheep or indeed having been any problem whatsoever. As the animal was described as very large, and the species displayed significant sexual dimorphism (Rovinsky *et al.* 2020), it is likely though not certain that it was a male, particularly as only one report of an outsized female appears to exist: a mother with four young caught in a snare measured 6 feet long with a short tail (*The North West Post* 1888), which may have been a product of bounty-era propaganda.

Given the prey size constraints of the thylacine (see pp. 20–3), we would expect that most reports of sheep predation would relate to lambs. This is borne out in the published sources, as there was only one reliable instance of thylacine predation upon an adult sheep prior to the first bounty, but four instances of lamb predation. The first known case occurred on Abbott's farm in late October 1821, when, at around 4.00 pm in the afternoon, a thylacine appeared and chased the flock downhill in view of a shepherd and killed a lamb (*Hobart Town Gazette and Van Diemen's Land Advertiser* 1821). Two years later, a second instance of lamb predation occurred at Coal River east of Hobart and the Derwent River, in which a small terrier famously killed his much larger foe (*Hobart Town Gazette and Van Diemen's Land Advertiser* 1823). Three years later, a thylacine was again killed near the Coal River and upon examination of its stomach contents, two recently eaten lambs were found (*The Australian* 1826). Although lamb predation was more common than predation upon adult sheep, it was still virtually unheard of.

Less than 1 month after the death of the latest lamb killer, a third thylacine was killed on Abbott's farm, though, as with the first thylacine killed, no mention was made of any danger to livestock or human inhabitants (*Hobart Town Gazette* 1826). Although there is no reference to it being a female, this is clearly evident from the fact that four young were found in the pouch. Abbott, perhaps sympathetic to the four defenceless young, or possibly just curious, attempted to raise them. This had ambivalent short-term success, as 'they are alive and healthy but show but little symptoms of domestication' (*Hobart Town Gazette* 1826), although their ultimate fate remains unknown, as he does not appear to have kept a diary and no other record of the small, striped quartet can be found.

Given the objective treatment of the species in the three newspaper articles relating to Abbott's farm, wholly without exaggeration or propaganda against the species, even though a lamb was killed and a flock worried, it is clear that the species was not universally hated by the agricultural community. Later, more successful attempts at keeping the species in captivity (Paddle 2000), further supports the idea that not every Tasmanian wanted to kill every thylacine on sight. Instead, the general attitude of people towards the species was more ambivalent and nuanced.

Edward Abbott III (1801–1869) would go on to be a well-known aristologist (the art or science of cooking and dining), who anonymously authored what has often been considered the earliest Australian cookbook (although published in England) in 1864 (Abbott 1864). If more people followed his recipe for treatment of the thylacine, then it would have produced more than food for thought. The species might be encountered in the wild today, instead of just in museums and private collections.

PART 5: THE BOUNTY YEARS (1830–1914)

Merino sheep and scapegoats: a bounty of human ignorance

Branden Holmes

Sometimes words that have little contemporary importance can retrospectively take on enormous significance: 'Although these animals are by far less numerous than the native dog in New South Wales, if the increase of them is not stopped in time, they may become formidable. We would suggest, then, that a reward of some description should be given by the Government to pay everyone who shall destroy and bring into town a native tyger or hyena' (*The Australian* 1826).

Having been isolated on the island of Tasmania for thousands of years, the thylacine was not experiencing a serious population increase in the early 1800s. In fact, the population was fairly stable. Yet 4 years later (1830) saw the introduction of the first bounty on the species by the Van Diemen's Land Company, despite only five published accounts of specific thylacine predation upon European livestock totalling one sheep and four lambs in the 27 years since European settlement. Only a few rare instances of sheep predation, and a handful of thylacine attacks on people, from provocation or out of desperation, would follow over the decades to come. By any serious measure, the thylacine remained a non-threat to humans and their livelihoods until its extinction.

Dogs on the other hand were a major problem. They were brought to Tasmania at first European settlement in 1803, but irresponsible owners soon let them roam free. In 1814, W Mitchell noted that several of his sheep had been killed or maimed by dogs (*The Van Diemen's Land Gazette and General Advertiser* 1814). In 1817, an infant was badly bitten by a ferocious dog while playing around the house, only the latest in a string of incidents (*The Hobart Town Gazette and Southern Reporter* 1817). And by 1819, roaming dogs were regularly attacking people in the streets (*The Hobart Town Gazette and Southern Reporter* 1819a).

The European genocide of the Palawa (Tasmania's Aboriginals), known as the Black War (1824–1831), contributed heavily to the increase in the feral dog population. The Palawa had prized canines since their introduction, even allegedly suckling them at the human breast, as they greatly helped them hunt kangaroos and other native animals (*The Hobart Town Courier* 1829b). When the scattered Palawa survivors of the genocide were taken to Flinders Island between 1831 and 1833, where most succumbed to illness and died, their dogs no longer had human companions. These now feral dogs turned to the unmitigated hunting of native game, including the Tasmanian emu (*Dromaius novaehollandiae diemenensis)* and sheep (*The Hobart Town Courier* 1829b). It was now not uncommon for shepherds to find feral dogs worrying a dozen sheep in a single morning (*The Hobart Town Courier* 1829a). As for the emu, with its restricted distribution, it soon became extinct, with the last confirmed sightings in the late

1830s (Dooley 2017). Yet, despite the overwhelming evidence pointing towards feral dogs, the propaganda campaign against the thylacine only intensified.

In 1833, Dr Temple Pearson advertised in the *Colonial Times* newspaper that he would shoot any dogs found wandering off the road near his property as they had killed many of his sheep. The newspaper's staff remarked upon the novelty of Pearson's intention to poison the carcasses with arsenic and suggested that the method might also kill thylacines and other native animals (*Colonial Times* 1833). In 1834, the importation of four beagles from England was welcomed as they could track the scent of vermin such as the thylacine, which are 'found to be so destructive to the flocks' (*Colonial Times* 1834). And in 1836, Edward Archer advertised his need for two dogs that were capable of killing thylacines and free from the vice of killing sheep (*Launceston Advertiser* 1836).

A second bounty by the Van Diemen's Land Company was introduced in 1840 after the first had ended in 1838 and ran until 1914 (Paddle 2000). The same year, John Waddlesworth was apprehended by police with a loaded pistol, claiming that he was carrying it for protection against thylacines, a story that his employer, Mr Jones, backed up, stating that it was not uncommon for shepherds to 'carry arms to protect themselves from the native tigers' (*Launceston Courier* 1840). In the second half of 1841, a further two bounty schemes came into existence, and probably both went out of existence. The third, short-lived, bounty on the thylacine by the Clyde Company ran from July to October (Paddle 2000) and a group of farmers offered £2 for any thylacine killed within 5 miles of any of their properties (*Colonial Times* 1841).

As a crepuscular or nocturnal species, and with a small European settler population, early thylacine encounters were very rare. Brief observations of captive specimens did not provide much knowledge and studies of the species in the wild were non-existent. So little was known about the species, and there was little inclination to learn, that propaganda did not first need to displace established fact before being accepted. In the first decade of the 19th century there were no negative portrayals of the species published. In the second decade there were five. During the third there were eight. And by the fourth there were fifteen, excluding reprints. The species was well on its way to extinction by this point, but it could also easily have been saved.

The 'Philosopher' and the thylacine

Nic Haygarth

James 'Philosopher' Smith (1827–1897), discoverer of the great tin lodes at Mount Bischoff, 70 km south of what became Burnie, had many encounters with thylacines during his solitary mineral prospecting expeditions in the north-western highlands. By 1862 Smith had learnt about the curiosity of the thylacine when it met humans in the bush. He stated that he had never known a Tasmanian tiger to attack a man, although he knew of a case where a thylacine walked up to two men in the bush and 'was knocked on the head before he had shown any aggressive intentions' (Forth 1862). He also knew of a case of a thylacine following a man through the bush at night, a behaviour reported by others in daylight hours (Forth 1862).

However, by the end of his bush career Smith was more wary of thylacines. It is easy to imagine that a man working alone in the trackless highlands, with no hope of rescue, would feel vulnerable in the presence of a large carnivore. 'I thought it necessary', he wrote: 'to be on my guard against them by keeping a fire as much as I could and having at hand a weapon with which to defend myself in the event of being attacked by one or more of them' (Smith unpublished).

He regarded the thylacine as the only threat to his survival in the bush, stating that:

> It's all very well to talk about not having any dread of tigers when you sleep in an enclosed tent, but when you sleep in the open air in fine weather as I did for years it would be an easy matter for a tiger to approach and seize you by the throat while you are lying asleep. Only once however did I see a tiger, a very large one, approaching while I was lying at my campfire … he was creeping towards me in a very stealthy manner, but on my springing to my feet, and shouting at him and menacing him with my axe, he went away (Smith no date).

On another occasion, while cooking food over a campfire in the Vale of Belvoir, Smith heard a thylacine grinding its teeth behind him. It was often said that thylacines were attracted by the smell of frying meat. Again, the raising of Smith's trusty axe caused a thylacine retreat. That night he made sure that his tent was 'well closed' when he retired (Smith no date).

It seems a common belief today that all dogs were scared of thylacines, but this was not the case. Dogs trained to hunt would willingly engage a thylacine. Smith had even seen a sheep dog kill one (Forth 1862). His own dogs were not formidable staghounds or 'kangaroo dogs', but often domestic crossbreeds. On one trip he took a Skye terrier – a tiny, short-legged dog – which guarded his knapsack when he was camped at the Leven River. Smith made a fire beside the river, but while he was trying to land blackfish for his dinner the dog began to bark furiously. He saw it tearing towards him with a thylacine in pursuit. The prospector heaved a stone, which hit the pursuer in the side. The thylacine 'checked his speed a moment and, turning his head, snapped his jaws in the direction in which he had been struck'. Smith had made a habit of carrying a revolver with him while at the Victorian gold rushes of the 1850s and he continued to do so in the Tasmanian bush. Grabbing the revolver from his knapsack he fired and wounded the tiger as it continued to chase his dog. It finally escaped into some dense scrub (Smith no date).

Smith came to regard what he called Tiger Plain, on the northern edge of the Lea River, as the most thylacine-infested place he visited during his expeditions. It was popular among the carnivores, he thought, because by driving game in that direction a Middlesex stockman's dogs provided a ready food source.

It was here that with the help of his dog Smith killed the largest thylacine he had seen at that time, its head and body measuring 3 feet 4 inches, with a 17-inch tail (Forth 1862). It was a female with four young at the teat inside her pouch, which might explain the ferocity of the thylacine's defence when engaged by Smith's dog. She continued to fight the dog even when Smith fractured her skull with his tomahawk and it was only when, having grabbed her by the hindlegs, he was able to sever her spine that the animal died. Then Smith discovered the pups, which were just old enough to walk. Like many 19th century bushmen, Smith was an amateur naturalist. He considered keeping the young as 'specimens of animated nature', but found of course that they could not digest prospector food, damper and bacon, or blackfish or any native animal meat he could get (Forth 1862).

Smith killed a smaller thylacine on the northern side of the Black Bluff Range in similar circumstances, knocking it on the head when it was engaged fighting his dog. He believed he was justified in 'destroying one of a kind which is often very destructive to sheep and lambs' (Forth 1862). Unfortunately, Smith was far from alone in that belief.

Thylacines in European zoos

Stephen R. Sleightholme and Cameron R. Campbell

The history of thylacines exhibited in European zoos is relatively well documented. A total of nine thylacines were displayed in four continental collections (Berlin (4), Paris (2), Antwerp (1) and Cologne (2)) from 1864 to 1914. Other specimens are known to have been dispatched to Antwerp and Paris, but they all perished in transit.

The Zoologischer Garten in Berlin was the first zoo in Germany, opening its doors to the public on 1 August 1844. A total of four thylacines were exhibited at the zoo between 1864 and 1908. The first of Berlin's thylacines, a male, was purchased from the Regent's Park Zoo in London. It arrived at the zoo on 7 June 1864 and was placed on display in the small carnivore dens. Unfortunately, its stay was short-lived, the animal dying on 14 November of the same year. Moeller (1997) notes that this thylacine holds the longevity record for any thylacine in captivity at 8 years and 220 days. Upon its death, its body was forwarded to the Museum für Naturkunde (Humboldt University), where it remains in the collection to this day as specimen ZMB 2986 (Sleightholme and Ayliffe 2017).

A second thylacine was purchased by the zoo's Director, Karl Bodinus, from an unknown source in 1871. It was placed on display in the large carnivore house (Raubthiere) until its death in 1873. The last of Berlin's thylacines were an adult pair, purchased from the Reiche brothers of Alfeld (via H. E. August Görling in Sydney), arriving on 25 February 1902. The female died on 23 December 1905 and the male 3 years later on 16 January 1908. The male is the only thylacine in continental Europe for which a photograph still survives (Fig. 10) (Sleightholme and Campbell 2021a).

Ludwig Heck (1912), the Director of the zoo from 1888 until 1931, made the following observations on the zoo's thylacines:

> They act quite familiar coming restlessly up to the cage bars and sniffing around if one stands on this side of the barrier directly in front of the cage.

Fig. 10. Male thylacine on display in Berlin Zoo (*c.*1905). Photo courtesy: Moeller archives.

> Fired by eternal greed, they constantly demand food when they are not sleeping.... They keep trying to chew through the bars. They are hard to arouse from their sleep on their soft straw beds in the dim night cage, but are not unpleasant if you do awaken them... otherwise they pace for hours in the cage without paying much attention to the outside world, or lie quietly, sleeping apathetically.

Some of the traits observed by Heck are now known to be stereotyped behaviours exhibited by stressed captive animals. For example, several of the skulls from the Berlin thylacines, now housed in the Museum für Naturkunde (Humboldt University), show significant damage to the canine teeth caused by the animals repeatedly mouthing the bars of their cage.

Founded in 1873 from the Royal collection, the Ménagerie du Jardin des Plantes in Paris is one of the world's oldest public zoos. Two thylacines, an adult pair, were displayed in the Predator House (Fauverie) between 1886 and 1891. Both thylacines were obtained by Albert Le Souëf, the Director of Melbourne Zoo, from William McGowan of the City Park Zoo in Launceston. They arrived in Melbourne on 19 January 1884 and their addition to the collection noted in *The Age* newspaper on 6 February 1884:

> A pair of marsupial wolves (Thylacinus cynocephalus), from Tasmania, had also just been added, they being of special value, owing to the difficulty in procuring them (*The Age* 1884).

They remained on display in Melbourne for 2 years, before embarking on a voyage to London with Dudley Le Souëf, the Assistant Director of the zoo and his son, in January 1886. The thylacines arrived at the Regent's Park Zoo on 19 March 1886 and their exchange

Fig. 11. Thylacine enclosure at Cologne Zoo. Photo courtesy: Moeller archives.

was agreed with the Ménagerie du Jardin des Plantes for £90 of stock animals. They departed for Paris the following month on 15 April 1886. In a report written in 1891 by Alphonse Milne-Edwards, the zoo's Director of Mammals & Birds, he stated that the 1890–1891 Parisian winter was especially cold and that among the total of dead animals were the zoo's two thylacines (Milne-Edwards 1891).

Virtually no records survive for the two male thylacines that were displayed at the Cologne Zoo between 1903 and 1910 (Fig. 11). It is known that both were purchased from Carl Reiche, a rival of the German animal dealer and zoo proprietor Carl Hagenbeck. Reiche obtained the animals from the Sydney-based animal dealer H. E. August Görling. The first thylacine arrived at the zoo on 26 March 1903 and cost 300 marks; the second arrived at the zoo 2 months later on 13 May 1903 and cost 800 marks. No reason is given for the disparity in pricing of the two specimens. Following their deaths, their bodies were sent to the Cologne Museum of Natural History (Staplehaus), but regrettably, the museum was destroyed by Allied bombing in the war and the specimens and their accompanying records lost. The first of the zoo's males to die was on 27 September 1909 and the second on 13 May 1910.

Antwerp Zoo in Belgium was established in 1843 and is one of the oldest zoos in the world. The first of the zoo's thylacines was purchased from Melbourne Zoo in 1903, but died during transit. The second of the zoo's thylacines, a young male, was purchased by the animal dealer John Hooks from fellow dealer James Harrison of Wynyard (Tasmania). It transited through Melbourne Zoo, finally arriving in Antwerp on 6 February 1912, where it was housed in the small carnivore gallery (Fig. 12) until its death on 13 February 1914. After its death, its body was donated to the Institut Royal des Sciences Naturelles de Belgique, where it remains in the collection as a disarticulated skeleton (IRSNB 32B) (Sleightholme and Ayliffe 2017). The death of the Antwerp thylacine marked the termination of thylacine displays on the continent, although the Regent's Park Zoo in London continued to display the species until 1931.

Fig. 12. Small carnivore gallery at Antwerp Zoo. Photo courtesy: Eric Block.

Never far apart: picturing Paris's pair of pouched predators

Branden Holmes

It is incredible that any live thylacines were displayed outside of Australia given the long, cramped sea voyages they invariably endured. A female (concealing pouch young) sent to Washington Zoo travelled for months in a box that was so small she couldn't even turn around and it is suspected that at least seven thylacines were sent to Paris's official menagerie at the Jardin des Plantes (*The Age* 1863; Paddle 1996), yet the zoo records only two animals arriving alive (Moeller 1997).

This appalling survival rate would have been even worse if not for a remarkably long-lived and well-travelled pair of thylacines. Spanning almost 8 full years, excluding their unknown age at capture, they are among the best documented living specimens ever recorded. Collectively, various newspaper articles, official zoo records and visitor accounts tell the story of an inseparable pair caught in colonial Tasmania in mid-1883, with short periods in the care of Dr Crowther (several months), Melbourne Zoo (2 years) and London Zoo (1 month). Their final 5 years were spent at Paris's Jardin des Plantes, dying less than 2 months apart in the bitter Parisian winter of 1890–91.

They were caught by Mr A. W. Brewer near Bridport in July or August 1883, said by him to be the only pair taken alive out of 60 animals (*Launceston Examiner* 1883a), which if true, would make him one of the most prolific of all thylacine trappers. Dr Arthur Bingham Crowther subsequently purchased them to fulfil a persistent request by Sir William Henry Flower, Director of the Natural History Museum, as specimens for the London Zoo (*Launceston Examiner* 1883b). However, the sale proved problematic and Crowther eventually sold them to the Melbourne Zoo instead (*Launceston Examiner* 1884). They arrived safely (*The Argus* 1884) and went on display to visitors (Wyuno 1884) for 2 years before being exchanged with the Jardin des Plantes for £90 worth of animals. En route to Paris, they briefly stayed at London Zoo, thus partially fulfilling Dr Crowther's original intent.

According to the zoo's entry book they arrived at the Jardin des Plantes on 17 April 1886 and were housed in the Predator House (Fauverie), which has long since been demolished. Sadly, little written description or visual depiction of their quarters exists, though Friedel (1890) speaks of having observed them twice, both times in the sunlight and another visitor's account provides specific details about their diet, sometimes being fed live prey including guinea pigs (Huet 1887). Huet also mentions two other very interesting observations: that the animals never vocalised even when excited such as by the proximity of a dog and were quite playful with their keeper.

Tasmania frequently reaches subzero temperatures, so theoretically, the thylacines would have been better adapted to cope with the particularly bad winter of 1890–91 than other animals at the zoo. The fact that they both succumbed within a short period suggests that, at a relatively advanced age compared with the longevity of other captive specimens, they had reached the end of their expected lifespan. According to the zoo's official exit book the female died on 6 February 1891 and her male companion died 53 days later on the last day of March.

Today, their physical remains are preserved as specimens MNHN 1891-61 [♀] and MNHN 1891-327 [♂] in the Muséum National d'Histoire Naturelle in Paris (Sleightholme and Ayliffe 2017). Despite being housed together for almost 8 years they failed to reproduce, so either they were siblings or the conditions under which they and many others of their kind were kept were too unnatural to be conducive to breeding. With the exception of the brief period the male remained alive after the female had died, for more than 137 years, they have been, and remain, together.

The myths of the thylacine hunter and of a successful campaign of extermination waged against the thylacine

Nic Haygarth

The thylacine (or Tasmanian tiger) hunter is a product of romance, with no basis in reality. The idea that thylacines were decimated as a vendetta perpetrated by woolgrowers is equally romantic. The demise of the thylacine during the 19th and early 20th centuries was not the result of systematic persecution of the animal, but rather was incidental to the development of the pastoral industry, the spread of rural settlement and of the mining industry in Tasmania with an epidemic disease finally decimating a population already at dangerously low levels.

Most thylacine killings were unwitting or opportunistic, rather than part of a concerted campaign to eliminate the animal or collect bounty payments. These killings happened largely in the pursuit of kangaroo, wallaby, pademelon, ringtail and brush possum furs, a pursuit that went on regardless of whether bounty payments were available for killing thylacines. In other words, the thylacine was 'collateral damage' in the Europeanisation of Tasmania.

Thylacines lost much of their available food source and much of their habitat during the first century of European settlement in Van Diemen's Land (as Tasmania was known until 1856). They were displaced by the clearing and occupation of land and threatened by the introduction of the dog, competition for game and the widespread hunting of native animals. The spread of pastoralism on the thylacine's hunting grounds put it on a collision course with the stockman–hunter; the man paid £20–30 annually by the pastoralist to protect stock from predation, theft and disease. The duties of the stockman–hunter left plenty of time for hunting and, with a team of dogs, he could lay waste to the native animal population, providing him with meat and extra income from saleable and tradable furs.

On some pastoral properties, such as at Woolnorth in the far north-west, hunting the adjoining country with dogs was so much more lucrative than tending sheep and mending fences that often it was difficult to keep men at their posts. Kangaroo hunter's dogs were a menace to sheep, but outside the realm of the big woolgrower there was unfettered dogging. This was the hunting method of choice for highland stockmen such as James Lucas and Charlie Drury on the Surrey Hills near latter-day Waratah and Harry Stanley on the upper Forth and Mersey River systems. Staghounds trained to kill would swiftly dispatch any thylacine whose scent came to their attention. Many thylacines must have been torn limb from limb by powerful dogs like these. Dogs that went wild would also have become competition for the thylacine's food source of smaller native animals.

The bush farmer was another menace to the thylacine. This was the small land owner who carved a farm out of the forests long after the best grazing land had been alienated by woolgrowers. Like shepherds and stockmen, virtually all Tasmanian bush farmers were hunters and some, such as Jerry Aylett of Elizabeth Town, joined highland stockmen in pursuing the thick winter pelts, especially brush possum pelts, sought by skin buyers for the export trade. Different hunting techniques were used for different animals. Possums were shot at night; brush possums were also snared in pole snares and ringtails trapped in rabbit traps or poked out of their nests and shot in daylight. Dogs were used to pursue kangaroos, wallabies and pademelons, but these marsupials were also frequently snared. As suggested by Paddle (2000), strychnine baits used by snarers to kill off scavenging Tasmanian devils may also have been taken by thylacines.

From the second half of the 19th century onward, prospectors and miners were blamed for driving thylacines out of their usual habitats and onto pastoral runs. Typical mining impacts of large-scale burning to reveal mineral outcrops and to improve access, deforestation, pollution of waterways, the establishment of new settlements and the introduction of dogs would have been disastrous for the thylacine. Some hunter–prospectors, such as Charlie Davey and William Aylett, lived in the bush all year round, hunting on a daily basis and relying on kangaroo and wallaby for meat. Their companion dogs were trained to hunt. The carnivorous thylacine was never considered a food source, but inevitably, prospectors and miners drove it from its habitat, competed with it for food, shot it opportunistically and snared it unwittingly.

The necker snare, made of steel wire, brass wire or hemp, was placed at a height that caught the animal in a noose, breaking its neck or strangling it. Thylacines were known to escape treadle or footer snares, which caught the animal by the paw, but were often killed in necker snares intended for other animals. Police Sergeant M. A. Summers (unpublished), who conducted thylacine searches for the Scenery Preservation Board in the 1930s, described the necker snare as deadly and 'the real menace' to the thylacine. On the other hand, Summers asserted, the springer footer snare had no effect on the thylacine.

This susceptibility to the necker snare is borne out by its use by shepherds such as Joseph Willett, Alex Coplestone (or Copplestone) and the Woolnorth tigermen who deliberately set out to catch thylacines as part of their job of protecting their masters' flock. Willett, the shepherd at 'The Island', on the Macquarie River near Tooms Lake, claimed to have captured more than 20 thylacines, mostly in necker snares set on their runs or in gaps in fences. North-eastern stockman–hunter Alex Coplestone killed 15 thylacines in amost 3 months in necker snares made of 12 strands of twine (*Launceston Examiner* 1897; Coplestone 1926). At Green Point on the Woolnorth estate, a line of necker snares was maintained to catch thylacines entering the property.

However, the occasional deliberate hunting of thylacines should not be confused with being a specialised or professional thylacine hunter. Men such as Willett, Coplestone, Stevenson and the Woolnorth tigermen were fairly typical stockman–hunters who received a small wage, which gave them an incentive to participate in the fur industry by killing the animals that competed for feed with their master's stock. Although these men profited by killing thylacines, their main source of income would have been from the much more plentiful brush and ringtail possum, kangaroo, pademelon and Bennett's wallaby skins.

The suggestion of there being specialist thylacine hunters is simply a misunderstanding of the Tasmanian fur industry. There was little financial incentive to hunt thylacines. The thylacine was not hunted for the value of its fur and appears to have existed in much smaller numbers than the animals that were valued. About 2184 thylacine carcasses were submitted for payment under the government thylacine bounty scheme in the years 1888 to 1909, an average of about 100 per year. In 1888 alone 173 882 possum, kangaroo, wallaby and pademelon skins were shipped from Launceston to Melbourne and 280 016 in 1889, not taking into consideration skins exported through other ports or those sold locally. In 1903, 285 000 native animal skins were exported from Tasmania. Why would hunters who had a ready market for skins available in such numbers spend their time trying to hunt down the comparatively scarce thylacine for the sake of a £1 bounty?

The idea of hunters having a commercial imperative to kill 'tigers' seems to hinge upon mostly speculated payments by private bounty schemes and upon Norman Laird's claim that he possessed a record of 3482 thylacine skins being shipped to London to be made into waistcoats in the years 1878–96 (Laird 1968).

First, to deal with the private bounty schemes. Guiler and Godard (1998) suggested that private and government thylacine bounties enabled some men to make a living from killing thylacines alone. However, no evidence was given to support this claim, although they did mention an assertion that the Oatlands–Ross Landowners Association paid out about 40 bounties of £5/10/0, one shepherd alone collecting 18 bounties (Guiler 1985). Yet because he was a shepherd, this bounty collector was clearly not a dedicated thylacine hunter, and although he would have collected a handy £99 from the Oatlands–Ross Landowners Association, we do not know the period over which these animals were taken. Perhaps the bounties were claimed over a 10-year period. This is clearly an example of a stockman–hunter receiving additional income from the incidental killing of thylacines. The highest individual tally for a hunter on the government thylacine bounty list appears to be James Pearce's 52 bounties (50 adults and 2 juveniles), collected over a period of 11 years and worth £51, an average of about £4/13/0 per year. This was no more than a useful supplement for skilled snarers who made most of their income from an open hunting season.

The best documented thylacine bounty scheme is that which operated on the Van Diemen's Land Company's Woolnorth property from 1871 to 1912, although, as Robert Paddle has asserted, the so-called Woolnorth tigermen were simply stockmen who occasionally killed thylacines (Paddle 2000). The Company's records show that it paid out thylacine bounties of about £121 during this period, or an average of about £3 per year – pocket money for a man with a salary of £20–30 and probably a greater annual income from the hunting industry. In April 1879, for example, tigerman William Forward sent 180 wallaby and 84 pademelon skins to market (Wilson unpublished). In 1879 the open season for 'kangaroo' (wallaby) was from 31 January to 31 July, suggesting that Forward's haul represented at most 2 months' worth out of a 6-month season. By market prices of the time, these skins would have been worth at least £8/4/0 and at most £14/14/0 (*Cornwall Chronicle* 1879; *Launceston Examiner* 1879).

Second, to deal with Norman Laird's claim that thylacine skins were exported to London, where they were turned into waistcoats. There are no shipping, customs or skin buyer advertisements for thylacine skins to support this claim, which was debunked by Paddle (2008). Paddle suggested that Laird's thylacine skins were more likely dingo skins exported to London as 'Australian wolf'.

A certain amount of 'tiger farming' went on in Tasmania when the animal became more valuable alive than dead. In 1890 a live wether might fetch 13 shillings, compared with £6 for a thylacine delivered alive to Sydney. At this time bush farmers such as Joseph Clifford and Robert Stevenson effectively added tiger farming to their portfolios of raising stock, growing crops and hunting. Because thylacines already appeared on their property, they made a conscious effort to catch them alive in a foot snare or pit rather than dead in a necker snare (H. Clifford, *pers. comm.* to D. Groves). This was not tiger hunting. 'We didn't go shooting tigers', Harry Clifford, son of Joseph Clifford, asserted. 'They came to us'; that is, they entered the Clifford property and ended up in Clifford snares (H. Clifford *pers. comm.* to D. Groves)

Elias Churchill, the rural timber worker celebrated by some as the captor of the last thylacine to die in captivity, never spoke publicly about the thousands of brush and ringtail possums, wallabies and pademelons he would have killed while snaring a very small number of thylacines – because nobody asked him. It wasn't romantic enough. Churchill told Michael

Sharland (1957) that the thylacines he caught in the Florentine Valley 'blundered into his snares', a far cry from them being bailed up during tiger safaris. Other general hunters such as Wally Mullins, Harry Loveluck and Paddy Hartnett caught thylacines alive in the same area at about the same time, yet nobody hailed them as thylacine hunters. In the late 1920s, when Wynyard animal trader and sometime snarer James Harrison sold live thylacines for up to £75 each, he staged expeditions to the Arthur River specifically to secure tigers, resulting in at least one success in 1929. Of all those who caught thylacines alive up until 1931, perhaps Harrison alone, at this late stage of the animal's demise, could lay claim to the title of 'tiger hunter', yet we know that he bought the vast majority of the living thylacines he traded from prospectors, general hunters, bush farmers and their families. Tiger hunting, with its suggestion of pith helmets, safari suits and mounted trophies, belongs to the British Raj in India, not in the Tasmanian bush.

Thylacine capture site at Meadstone: a tiger lair or sunny resting place?

Kathryn Medlock

The thylacine is currently a symbol of the appalling devastation caused by humans to the natural environment. With extinction came interest, which in turn has led to increased research, a trend that continues to this day. Of necessity, current research revolves around physical objects such as specimens in museums, and related material such as photographs, historical records and oral history. This essay is a first attempt to focus on the importance of place in the thylacine historical record.

Some sites that relate to thylacine history are known and recorded. For example, in Hobart, both the Battery Point and Queen's Domain sites of the Beaumaris Zoo, where thylacines were kept in captivity between 1908 and 1936, are well recorded. The Domain site in Hobart has become an iconic site as the location of the death of the last known thylacine that died there in September 1936. Although the thylacine cage no longer exists, the gates are decorated with thylacine imagery. The heritage value of this well-known site is clearly recognised.

Other sites of significance to thylacine history are less well known. For example, the names and locations of farms where thylacines were killed as pests are recorded in history, but do not feature as particularly significant locations.

The purpose of this essay is to document the site of a wild thylacine capture near the St Paul's River in north-east Tasmania. This site is the only recorded exact location of a wild thylacine capture from the natural environment as opposed to those that occurred in an unspecified general area or on farms.

The capture site

In 1902, the Director of the Royal Melbourne Zoological and Acclimatisation Society Mr W. H. Dudley Le Souëf was approached by the director of Antwerp Zoological Gardens in Belgium, F. Krueger, asking for a thylacine. With no suitable thylacines in stock, Le Souëf suggested that he travel to Tasmania to capture a living thylacine for Antwerp. To justify the expense of the trip, he suggested that he could use the trip to promote the work of the Zoological Society by presenting a series of public lectures and a magic lantern show. Through this means he hoped to encourage more donations to the Zoological Society.

While in Tasmania, Le Souëf presented several of his popular natural history lectures across the north-east of the island before taking a break to visit his friends, the Franks, on their property Meadstone near the St Paul's River. While he was at Meadstone, a live thylacine was captured by the property's farm hands, who were paid £7 for the animal. There is no record of the method of capture.

The details of this expedition and the capture of a thylacine during the trip are recorded in detail by Paddle (1992) who discovered a photograph captioned 'Lair of Tasmanian Marsupial Wolf. Meadstone' in an album compiled by Le Souëf and located in the LaTrobe Collection of the State Library of Victoria (Fig. 13).

The photograph depicts a shallow cave in sandstone with ferns growing near the entrance. The floor of the cave contains leaf litter that Paddle described as thylacine nesting material. This photograph is the only contemporaneous image that pinpoints a wild thylacine capture site.

Fig. 13. Meadstone thylacine lair, 1902. Photographer: W. H. Dudley Le Souëf. State Library of Victoria.

Writing in 1907, Le Souëf described the area around Meadstone as ideal thylacine habitat, noting that:

> ... some portions of this country, especially on the hill tops and in the gorges, are very rough, and great masses of broken basalt lie about in rugged profusion, and it is in this wild country that the Marsupial Wolf and Tasmanian Devil have their homes. The former are now getting scarce, as every man's hand is against them, for they destroy sheep when they get an opportunity; they are now found only in the roughest and most densely timbered country, and may probably maintain a foothold there for some years to come. Their camp is generally in some hollow in a rocky cliff, or under some of the huge boulders, or in a hollow log, and they rest there most of the day. Their food consists of any bush animals they can catch, such as Wallabies, Rats, Kangaroos, &c. (Le Souëf 1907).

The thylacine

On his return to Melbourne, Le Souëf reported to the Zoological Society that he had been able to obtain a 'pair of Tas. Devils, 2 Black opossums besides a Tas. Wolf' (Paddle 1992). The thylacine was offered and then sold to Antwerp Zoo for £20, thus justifying the expense of the trip.

In March 1903, after a 2 month stay in Melbourne, the living thylacine was loaded aboard the GMS *Oldenberg* bound for Antwerp, Belgium. The *Oldenberg* was hailed as one of the fastest German mail ships and therefore a good choice for transporting living animals on the long trip from Australia to Europe. Unfortunately, the thylacine never arrived at its destination (Paddle 2012). Its fate has not been recorded, but it is likely that it died during the voyage.

The site today

In 2001, a team from the Tasmanian Museum and Art Gallery found the capture site on private property between Pretty Hills and St Paul's River in north-east Tasmania. The area certainly conforms to the description of thylacine habitat. It is a steep and very rocky area covered with sparse eucalypt forest.

The opening of the 'lair' measures 2.7 m wide and 1.2 m high and is located at the base of a sandstone escarpment. It is quite shallow with a depth of approximately 1.3 m. In front of the opening is a relatively flat area, which then falls away to a steep fern-covered slope. After 99 years, the site still looked exactly as it did when Le Souëf snapped his now famous photograph (Plate 11).

Approximately 6 m from this location, in a northerly direction, we found four smaller tunnels and several shallow cavities along the base of the escarpment. Two tunnels extended for more than 2 m into the cliff and are most likely used as wombat burrows or devil dens. There were numerous wombat scats around the entrances and one contained a wombat skull and other bones. The area above and below the escarpment is also littered with wombat scats, evidence of a large wombat population in the area. The shallower cavities are unlikely to be dens as they are too shallow to provide significant protection. They contained dried ferns, twigs and grass, most likely blown in by prevailing winds.

The 'lair' also contained dried leaves, grass and twigs that appeared to have blown in. Tasmanian devil and other small mammal scats were deposited on the flat area fronting the site. We determined that this was a Tasmanian devil latrine, which are where the normally solitary devils deposit scats to communicate their presence, like a calling card, to other devils in the vicinity.

The escarpment continues in a southerly direction for a further 15 m but contains no further tunnels.

Other thylacines on Meadstone

Just prior to this expedition, I met with an elderly man who had lived at Meadstone. He recalled that the old house on the property (now derelict) had been built before 1840 and that a stockade, for the protection and corralling of livestock, was built nearby. Anecdotal information is that in about 1912–14, six or seven thylacines were found in the stockade among the sheep. They were shot. This event has not been confirmed, but a newspaper article makes mention of a 'quite respectable number' of thylacines being removed from the area, crediting the 'prowess of the St Paul's sportsmen' for their contribution in reducing pests in the area (*The Examiner* 1907).

Conclusion

It was determined that the site was unlikely to be any kind of permanent lair or den in which a thylacine could raise its young or remain hidden during the day. It would not provide adequate protection from weather and predators (human or otherwise). It is shallow, open to the elements and does not lead to any inner shelter. It was, however, concluded, that the site was very likely to be where a thylacine could rest in the warm sun (it is north-facing) and then move elsewhere for cover when needed.

It is encouraging and nice to think that this site remains undisturbed and in its original state. Its preservation is, in part, due to the steep and rocky nature of the terrain. However, although it still looks the same as it did 99 years ago, it has clearly undergone significant changes. The largest apex predator, the thylacine, is gone. The second largest native Tasmanian predator, the Tasmanian devil, has (since our visit in 2001) suffered dramatic population decline due to Tasmanian devil facial tumour disease.

Although landscapes may appear the same, both historical and more recent changes in species composition have enormous implications that are usually not immediately obvious. On the surface, the site is as it was, but the loss of the thylacine and recent decline of Tasmanian devils has undoubtedly irreparably altered the ecology of the area.

Acknowledgements

Dr David Pemberton and Brian Looker assisted with the field work to locate the thylacine capture site. The current owner of Meadstone kindly gave us access to the land and advised on probable locations.

Dilger's tiger

Tammy Gordon

> … I carried him 10 miles and I tell you he was a fair weight.
>
> – Alfred William Dilger

The Queen Victoria Museum and Art Gallery in Launceston has a thylacine taxidermy in its collection with one of the best and most detailed histories of any in the world. Known as 'Dilger's Tiger' (QVM.1.2006), it was acquired in May 1912 and was the last complete thylacine to be added to the collection (Maynard and Gordon 2014). Although it has been in the Museum for over 100 years, the story behind its acquisition was only uncovered relatively recently. As with many thylacines in museum collections, the original label held little information (Sleightholme and Campbell 2018), but searches of archival records and information provided by the Dilger family allowed a more complete story to emerge.

In the early 1800s thylacines had little monetary value apart from small rewards on offer from various bounty schemes for dead animals. In the late 1800s and early 1900s thylacines were becoming rare and, fearing the species would soon become extinct, there was increasing demand for specimens by overseas and interstate museums, universities, zoos and private collectors (Paddle 2000; Maynard and Gordon 2014). This demand fuelled a local export industry. Captured thylacines, dead or alive, could be sold to animal dealers within Tasmania who would then sell them on for a profit.

This resulted in many thylacines, alive and dead, being taken out of Tasmania and eventually ending up in interstate and overseas collections. Tasmanian museums, relying mainly on donations and with limited funding, were unable to compete with the prices offered by animal dealers and so received little in the way of specimens (Maynard and Gordon 2014). Added to this, Tasmanian museums frequently exchanged their Tasmanian specimens for what were considered more interesting exotic species from overseas. As a result, Tasmanian museums hold relatively few thylacine remains compared with overseas institutions. The few that remain are therefore of special importance.

On 24 May 1912 Launceston's *Examiner* newspaper reported the following news: 'Mr Dilger, of Myrtle Grange, Mathinna, caught in a snare a splendid specimen of the Tasmanian Tiger, length 5 ft 1 in. It was caught about a quarter of a mile from his house. He is sending it to the Launceston Museum' (*Examiner* 1912).

Myrtle Grange was a property owned by farmer Alfred William Dilger and situated in the foothills of Mt Young, just outside of the township of Mathinna in Tasmania's north-east (Plate 12). Dilger discovered the thylacine on or shortly before 22 May. It was a fully grown adult male, presumably dead when found. The capture was accidental; like many farmers Dilger supplemented his diet and income by snaring wallabies and small game. Many thylacines were caught this way.

Dilger realised the significance of his catch and contacted Queen Victoria Museum Curator H. H. Scott to offer the thylacine for sale. In his diary entry for 22 May, Scott writes 'Tas tiger under offer from Mathinna (by phone) offered 20 shillings and costs. Accepted!' (Maynard and Gordon 2014). Scott records later that he purchased the thylacine for the skull only but later had it mounted. Even so, the price offered is low for the time. In 1912 the average price offered for a dead thylacine was around £3–5. No wonder Scott added an exclamation mark to 'accepted'.

The thylacine was sent to Launceston but it did not arrive at the Museum. Due to a misunderstanding, Dilger had wrongly addressed it to the Curator/Manager of Launceston's City Park Zoo, William McGowan. The City Park Zoo displayed thylacines and like most zoos also traded in both live and dead animals. Both the Museum and the Zoo were owned and operated by the Launceston City Council and although the Museum was a 'not for profit' operation, the zoo was more of a commercial enterprise. It is now known that at least 66 thylacines passed through City Park Zoo between 1885 and 1925, making it the most prolific thylacine trader known to date. Thylacines were either sold directly to buyers or exchanged for other animals (Paddle 2000).

Scott became aware of the mistake just in time and on 25 May notes in his diary that he, 'Nearly lost it as McGowan was sending it to Melbourne' (Maynard and Gordon 2014). It seems McGowan thought the thylacine was for him and wasted no time in getting it ready for a buyer.

Scott sent the thylacine to taxidermist William Notman to be mounted. Originally from Scotland, Notman had numerous occupations as well as being an accomplished taxidermist. He prepared many specimens for the Museum and, remarkably, was still doing so as late as 1930, at the age of 98 years old (Maynard and Gordon 2014). He died in 1933 at the age of 101. X-rays of the mount, taken in 2003, show the skull and lower limb bones are present, supported by a simple wire armature.

A month after Scott secured the thylacine for the Museum, he received the following letter from Dilger (4 June 1912):

> Mr Scott
> dear [sic] sir
> re [sic] the tiger I sent to you or Mr McGowen [sic] there was a mistake made I asked for Mr McGowen at the Post Office but they told me that you were the Curator & Mr McGowen the manager so I addressed it to Mr. McGowen thinking you were both the same I had a letter from Mr McGowen explaining things to me he said that he would of willingly give me 3 pounds for it as it was the best tiger he had ever seen I was going to wire to him but it was late in the evening and I did not have much time so thought it would be best to get him on the telephone that is where the mistake was made I would ask you to do your best for me in trying to give me more for the tiger as Mr McGowan said he would see you and try to induce you to give me more for it I carried him 10 miles and I tell you he was a fair weight.
> I am yours faithfully A. W. Dilger

Despite Dilger's plea the Museum Committee refused to alter the price and 'held him to his contract'. Dilger must have felt more than a little dismayed at this decision. After all, he had carried the thylacine 10 miles to Mathinna (and it could have weighed 20 kg).

Just over 1 month after Dilger's thylacine was captured, Scott received another offer of a dead thylacine to be sold to 'where the best offer comes from'. This time Scott sent the request straight to McGowan, who promptly purchased it for £3 and sold it on to the South Australian Museum (Queen Victoria Museum and Art Gallery Archives).

According to archival records held at Queen Victoria Museum and Art Gallery, 1 year later, in the winter of 1913, 15-year-old Edward Dearing snared a female thylacine with three young just outside of Mathinna within 10 km of where Dilger's thylacine was caught. This time they were alive and according to Dearing were all sent to City Park Zoo. These were the last thylacines recorded from the area.

By 1912, the devastating effects of European settlement had pushed the thylacine to the brink of extinction. Without government intervention or protection, the species stood virtually no chance of recovery. The story of Dilger's tiger provides a glimpse into the commercial exploitation that was likely to have been the final nail in the thylacine's coffin.

PART 6: A RAPIDLY DISAPPEARING SPECIES (1915–1936)

Mary Grant Roberts and the first Beaumaris Zoo

Gareth Linnard

One of the principal characters in any account of the thylacine is Mary Grant Roberts (1841–1921). Her private zoo 'Beaumaris', at Battery Point in Hobart, was noted for its displays of thylacines and opened to the public for the first time in 1895. Mrs Roberts exhibited 15 thylacines in her collection of native animals from 1908 until her death on the 27 November 1921 (Campbell 1999; *The Mercury* 1921). Her collection of animals was gifted by her daughter Ida to the Hobart City Council (*The Mercury* 1922a, 1922c), which relocated the collection to Queens Domain, opening the gates of the new zoo to the public on 2 February 1923 (*The Mercury* 1923).

The Beaumaris Zoo at its Domain location is now infamous as the place of the last captive thylacine's death. At its Battery Point location, the Zoo is remembered for Roberts' valuable contribution to the understanding of marsupial carnivores. Though she did not breed the thylacine in captivity, she was the first to do so with the Tasmanian devil. *The Sydney Morning Herald* (1910) complimented her efforts in creating a unique zoo:

> In one little house there is a family of Tasmanian tigers - the only one in captivity in the world - and it certainly speaks well for the owner that she has reared these creatures and kept them alive and well for several years, when the big zoological gardens have failed in the attempt. The habits and natural surroundings of every creature are studied; and, as far as possible, they are housed accordingly. Tasmania owes more to this public-spirited woman than it realises. Scientists from other lands recognise the value of her work, and in her visitors book appear the names of most of the notable people who have passed through Hobart; for her gardens are open to all who care to see them, and visitors are always welcomed and generally shown round by the owner herself.

The first of Mrs Robert's 15 thylacines was acquired in 1908 and was one of five specimens sold to the Regent's Park Zoo in London between 1908 and 1914 (Campbell 1999). As a result of her work with indigenous species, Roberts became a corresponding member of the Royal Zoological Society of London. In 1911 and 1917 she sold two thylacines to New York's Bronx Zoo and a further specimen, captured by Bill O'May at Tyenna, to Taronga Zoo in Sydney in 1918. She is also known to have traded in specimens from animals that died at her zoo.The skeleton of one thylacine, an adult male, which was captured by Mr Bill Power near Tyenna and was on display at the Beaumaris Zoo from 12 August 1911 until its death on the 9 March 1915, was preserved by Mrs Roberts and donated to the Tasmanian Museum in Hobart in 1922.

In its incarnation on the Domain, Beaumaris Zoo was in competition with the Wynyard animal dealer James Harrison for the ever diminishing supply of thylacines (Lord 1929) and paid full purchase price for those sufficiently badly injured as to make the amputation of a limb necessary for the animal's survival (HCCRC 1923a), and for three-legged specimens (HCCRC 1924a). Roberts had been far more particular; in 1917 she returned her final thylacine to Harrison because it had an inflamed foot.

On 7 July 1909 Launceston's *Examiner* (1909) records the shipment of a female thylacine and her three pups from Woolnorth to Roberts:

> Calling at Woolnorth, enroute, the *Gladys* took aboard four live native tigers, as they are familiarly termed, although bearing no affinity to the feline family. The animals (slut and half-grown cubs) were enclosed in a strong cage, and attracted a number of visitors on Sunday last. The beasts were transhipped to the S.S. *Taoroa*, having been purchased by a Hobart resident for, it is said, the sum of £20; another convincing proof of the wealth of our underdeveloped natural resources.

The three pups in the Woolnorth family group were sold and exported to the London Zoo between 1910 and 1911. Unfortunately, one of the pups, a female, died in transit. This family group was the first of only two to be photographed.

Mary Roberts was a fastidious diarist and recorded the purchases, destinations and deaths of the thylacines she acquired. It is likely that her full legacy and the context into which her work should be placed will only come to light with further research.

A brief glance at the Tasmanian fur trapper's effect on thylacines after 1909

Col Bailey

The period after the end of the official bounty scheme (1909–36) brought with it new challenges for the seasonal fur trapper, considering that the Tasmanian tiger was now a rarely sighted animal, having been predominately decimated by the reprehensible Tasmanian Government Bounty Scheme 1888–1909. Seldom now did a trapper set out to intentionally trap or snare a thylacine and the few that were caught alive and survived the experience were sold on to Australian and overseas zoos or collections mainly through local animal traders. Many trappers fortunate enough to unwittingly snare a live tiger obtained a considerable monetary reward. For several this was incentive enough to devote their skills to capturing this much-maligned animal.

If a trapper was fortunate to rescue an uninjured tiger from their snare line their immediate task was to safely transport the animal to a secure holding pen and thence to market. Tyenna trapper Elias Churchill told me in 1969 of the difficulty involved in this. He was adept enough to trap several tigers alive and sell them for a profitable return, as well as catching several more too injured by the trap to save. Some trappers killed the tiger there and then in an effort to rid the area of what was considered, along with the Tasmanian devil, nothing more than an occupational pest. The devil and the tiger were so often reviled because of their ability to devastate a whole trap line.

Central Highland sheep graziers, the Pearce clan, situated at various Derwent Valley localities between Ouse and Derwent Bridge, were amply rewarded for their much-acclaimed ability to both shoot, snare and trap native tigers they claimed were killing sheep on their various holdings. According to 'Pearce logic', the only good tiger was a dead one and they accentuated this belief by killing and surrendering almost 80 tigers for reward during the government bounty days. The majority of these were taken in the King William Saddle area of the Central Highlands.

According to Herb Pearce, the carnage continued well after the bounty days. As late in fact as the 1930s and possibly beyond, he said when interviewed by leading thylacine researcher Eric Guiler in the mid-1950s. Pearce told Guiler of recently shooting a mother and cubs found annoying sheep on one of the families' grazing properties. On another occasion a local returned soldier from the Korean War in the early 1950s told of seeing tiger pelts hanging from a tree on a Pearce property at Clarence River. They were a reticent, stand-alone unit who didn't tolerate outsiders, especially government officials. As a result, they entered local folklore as a renegade outfit that lived entirely by their own wits and standards. Guiler was extremely privileged to be welcomed into their domain and were it not for him virtually nothing would be known of their tiger hunting exploits.

Where once the range of the Tasmanian fur trapper extended to wherever there was a road or track along which to transport his bounty to an available market, there were also certain out-of-the-way areas, so isolated from transport corridors that almost guaranteed the long-term safety of their primary inhabitants.

There exists today in south-western Tasmania a section of wild and rugged country that has seldom, if ever, been visited. It has been that way because of its remoteness and isolation and it is here that nature appears to have guaranteed the welfare of what could possibly be the last bastion of the thylacine. For this magnificent animal its battle for survival has been a long and relentless one, a desperate struggle to endure against overwhelming odds. Following European settlement in 1803, sheep were introduced and as the tiger's reputation plummeted its eventual destiny was almost guaranteed. There have been few reported sightings of the tiger in this particular area over the past 85 years except by several mining prospectors venturing off the beaten track. One of these provided me with an excellently detailed report about 15 years ago.

My introduction to this challengingly difficult section of south-west Tasmania came via a meeting with legendary Melaleuca tin miner Deny King in 1980. King told me he was confident tigers still roamed this region of the south-west and so convincing was his testimony and general knowledge of the region that shortly after I decided to mount a personal challenge into this wild and largely unknown land. I canoed to the western extremities of Hydro Tasmania's Lake Gordon and ventured inland for several kilometres where, upon being confronted with the harshness of the terrain, I decided that this was not an exercise to be taken lightly. A planned expedition with entry of personnel and equipment by helicopter to a base camp from where incursions could be made inland appeared to be the best way of tackling this challenging quest. To date no such undertaking has been made that I know of.

Following the outlawing of fur trapping in Tasmania during the 1980s, this once much used method could possibly be the only way to successfully catch a Tasmanian tiger. Thankfully today's community standards would not allow it and for that we must all be grateful.

James Harrison: the last of the tiger men

Cameron R. Campbell

An often-forgotten figure in the history of the thylacine in captivity is the Wynyard-based animal dealer James Harrison. Detailed examination of his notebook, surviving correspondence and contemporary Tasmanian newspaper reports reveal that he played a far greater role in the trade of thylacines than was previously known.

James Harrison, the eldest surviving son of Thomas and Ann Harrison, was born in Nowgong, East India (on route to Tasmania from England) in 1873 (Fig. 14). His family

Fig. 14. James Harrison in front of one of his outdoor cages for the temporary holding of animals prior to sale. Courtesy: Archives Office of Tasmania.

settled in Boat Harbour on the north-west coast of Tasmania, where they lived for 34 years. Harrison moved to the nearby town of Wynyard in 1907 and was resident there until his death in 1943. A real estate agent by profession, he was also a respected naturalist and from 1910, became Tasmania's principal dealer in wild animals.

Harrison traded in a diverse array of Tasmanian fauna (both live and dead), from thylacines and Tasmanian devils to tiger cats (quolls), bandicoots, kangaroos, wallabies, possums, platypus and a wide variety of birds. The exact number of thylacines that Harrison purchased and sold will probably never be known, as surviving records are incomplete, but a realistic estimate would be in the order of 25 live animals and 20 dead specimens. It is known that Harrison placed his first advertisement for thylacines (colloquially known as native 'tigers' or 'hyenas') in the local classified press in 1910, and was actively dealing in the species into the early 1930s. Harrison maintained a number of holding cages for his stock on the grounds of his property at Wynyard. Blackwell (unpublished) states:

> He kept tigers in heavy wooden cages with iron bars, and drop floors to facilitate moving them from one cage to another. They were easy to handle and were never known to attempt to chew their way out as were the Devils he kept. After a time, the tigers used to get very tame. He used a net to envelop them when transferring them from a large cage.

All of Harrison's thylacines were procured in north-west Tasmania, many by well-known bushmen. The likes of A. R. 'Dick' Rowe, Almer Saward, Alf Forward, Frank Upston, Dick 'Boy' Evans, Harry Wainwright, Ray 'Turk' Porteus, Roy Anderson, Tom Harman and Adye Jordan all supplied live thylacines to Harrison.

Harrison's notebook provides a rare insight into his business practices. He appears to have been meticulous in recording the details of his sales, purchases and travel arrangements, and kept a running balance of his day-to-day accounts. Harrison notes that he supplied Albert Le Souëf (Director of the Taronga Zoo in Sydney) with marsupial rats in spirit, with the intestines removed; there are also several references to skulls (unspecified) and skins (unspecified), indicating that in addition to live animals, he also supplied specimen material to order.

Within his home in Wynyard, Harrison had his own private museum, consisting of three rooms in which he proudly exhibited numerous mounts and skins of Tasmanian marsupials, including 15 thylacine skins and four mounted specimens.

Harrison not only had first-hand experience with thylacines in the wild, but also personally secured, or helped to secure, at least three live specimens. Harrison is known to have supplied the Hobart and Melbourne zoos with live specimens. He also holds the dubious distinction of purchasing and selling the remains of the last recorded wild thylacine killed in Tasmania, a male shot by Mawbanna farmer Wilf Batty, on 13 May 1930. Harrison failed to secure any further thylacines following the Batty purchase. He did, however, continue to try.

James Harrison died peacefully at the local hospital in Wynyard on 29 September 1943, aged 83 years. He will be remembered as the last of Tasmania's 'tiger' men.

Just a Tasmanian animal: how familiar was the thylacine?

Gareth Linnard

The alleged capture of a 'Tasmanian tiger' by Bert Maher, near Scottsdale, in March 1953 (*The Mercury* 1953a) has been much revisited in recent years. After it was examined by a Mr P. Linton, described as a bushman, the body of the animal Maher had trapped, and then killed, was confirmed to be that of a thylacine. However, upon further inspection by Fisheries and Game Inspector N. D. McIntyre, it was identified as a quoll (*The Mercury* 1953b). It has been suggested, that this represents a piece of misdirection on the part of the authorities. After all, the objections go, how could two experienced bushmen fail to distinguish between a thylacine and a quoll? The proponents of this position may be unaware of two points. Firstly, no thylacines had been recovered from north-east Tasmania for over 30 years and, although Linton cannot be traced, Maher was 28 years old in 1953. Secondly, an almost identical incident had occurred 16 years earlier, when *The Mercury* (1927a) had eagerly crowned a quoll as 'tiger for a day'.

In 1927, the osmiridium mining fields of the Adams River were within the sphere of Elias Churchill, and probably several other thylacine trappers, when on 25 May *The Mercury* reported: 'An Adamsfield correspondent writes: It is reported that a splendid young tiger has been caught at the Falls… It is understood that the young animal has been sent to the Zoo at Hobart.' Yet, on the following day, *The Mercury* (1927b) reported the matter again: 'when it arrived the Curator found that it was no tiger, but a tiger cat'. Significant errors were obviously being made, even during the period when thylacines were still being traded, but by whom? In this case, the press are unlikely culprits, as it was a local correspondent who submitted the story. Further, as the sale to the Beaumaris Zoo was arranged not by *The Mercury*, but by the animal's captors, had the zoo been expecting a thylacine, it was they who gave that impression. So, the error almost certainly arose among the residents of Adamsfield.

From the 1920s onward, with the exception of a few dead thylacines bought as skins or specimens, or sold alive to James Harrison and Beaumaris Zoo, the trends noted while researching the late history of the thylacine are probably not much different to those of collecting post-1936 evidence: relatively few contemporary primary reports and a great many more made retrospectively in the years afterwards, with few of either of category inspiring total confidence. Yet, anecdotally it appears to be the case that people living at a time and place from where thylacines were definitely recorded are *assumed* to have been familiar with the species, but is this expectation supported by evidence?

A voice from the 1920s and 1930s that is often heard today is that of Alison Reid, though as we rarely get more than truncated pieces of her testimony, it's not always easy to understand the context of what she was trying to say. From her testimony, the Tasmanian public's reaction to Beaumaris' thylacines appears to have been lukewarm, 'they were just a Tasmanian animal' (*The Tasmanian Tiger* 1996), giving the impression of the thylacine being viewed as an unremarkable species, though this does not quite agree with testimony from elsewhere. The very fact that in 1924 and 1930, two thylacine captors, Walter Mullins and

Dan Delphin, took their animals on tour, advertised their locations in the press and charged an entry fee demonstrates their awareness that thylacines would be both popular and novel enough among their fellow Tasmanians to make their time and expense in doing so profitable. The editor of Hobart's *World* newspaper (1924) noted the popularity of the Mullins group, a mother and three cubs, at the city's Regatta: 'Despite the fact that the Hobart Zoo contains two fine specimens of the Tasmanian tiger, the sign "Wild Animals from Van Diemen's Land, Tasmanian Tigers", attracted a steady stream of patrons, who gladly parted with the nimble shilling to gaze on the products of their native isle.'

Even people resident in areas that were still definitely producing thylacines seem to have been eager to catch sight of the species. As was the case with Edith Gossage (née Marriott), of National Park in the Derwent Valley, who found it worthwhile to board the Fitzgerald to Hobart train in 1923 to be 'allowed to take a peep' at Churchill's thylacine in the guard's van (Gossage 1981). Adye Jordan (1987) wrote of transporting his thylacine in 1928 or 1929, there being some confusion in his account regarding the year, that:

> Many residents on the coast clearly remember the next Tiger from the Arthur River – it travelled from Takone (to) Wynyard in Mr Bill Bugg's passenger bus, it created tremendous interest, I was with it. Its passage to Wynyard was well known many days in advance and its passage was continually interrupted by sightseers and reached its destination much later than expected (Jordan 1987).

Paddle (2000) cites a reference from Doherty (1977) to the reaction of Lapoinya's residents to a capture in 1916: 'By nine that night, the bush telegraph had summoned from far and near, those who had or had not seen a thylacine alive. For miles the roads were blocked – phaetons, gigs, traps, buggies, jinkers and spring drays, hacks and bicycles and almost as wondrous as the tiger, there was a motor car'. Even with the suspiciously exuberant reaction at Lapoinya aside, contrary to Alison Reid's statement, thylacines appear to have pulled a crowd and it is important to ask why.

A certain degree of the novelty may have come from notoriety; but even then without unfamiliarity, necessarily born of the species' rarity, notoriety alone is unlikely to have been a major draw factor. The very fact that many of the relatively few witnesses from thylacine strongholds later came forward with only a single account of sighting the species during the 1920s tends to support this scarcity (Sleightholme and Campbell 2016), unsurprisingly so, with only 34 thylacine kills and captures contemporaneously recorded during the 1920s and 1930s.

Although there were Tasmanians with unique and intimate insights into the species, with a raw total of 147 kill, capture or sighting accounts from the 1920s and 1930s collated by Sleightholme and Campbell (2016) and visitor numbers at Beaumaris Zoo during the first 5 years of its operation at 501 705 (*The Mercury* 1928a), the majority of thylacine sightings of the era took place at a zoo. Following this line of reasoning creates a notable paradox. The Yearbook of London Zoo records that on the day of their last thylacine's death, 9 August 1931, the zoo received 4780 visitors; given the relative sizes of the populations of London and

Hobart, over time it seems all but inevitable that the majority of thylacine sightings during the 1920s took place in London's Regent's Park. Almost all of London's eyewitnesses would doubtlessly have been passive observers, though while the thylacine's cultural significance was already well established by the 1920s, the same is probably also true of many of witnesses in the species' home state of Tasmania.

Other Tasmanians of the era would only have seen the species through illustrations in the media. That said, between 1920 and 1940 the Tasmanian newspapers published only two separate photographs of the thylacine. Today, television, multiple publications on the species and the internet are a readily available reference source (Fig. 15) and above all, although

Fig. 15. Two thylacines at Beaumaris Zoo, Hobart. Tasmanian Archives: collection of photographs, PH40-1-3536.

many private images of the species have doubtlessly disappeared in the intervening years, 113 are known to have survived and have been collated in the 'Thylacine Image Registry' (Sleightholme and Campbell 2021a), with a further seven having been located in the past 2 years.

Modern Australians will be more visually familiar with the thylacine than many Tasmanians of the era, and not just urban Tasmanians. Sharland (1962) noted of a group of seasoned prospectors he had encountered in 1938, at the Jane River, between the species' two final strongholds, all of whom he described as experienced in the bush, 'none had ever caught sight of the animal, here or anywhere else'.

My father talked about it, but I didn't see it: the evidence

Gareth Linnard

It was noted by Rovinsky *et al.* (2020) that the thylacine's body size had become exaggerated in the literature. Their conclusion clearly highlights the risks involved in an incautious acceptance of questionable historical testimonies; by doing so, we are creating a false impression of the species. Without any confirmable first-hand experience of the species' survival beyond 1936, the more we permit our already limited historical understanding to become distorted, and for any resulting errors to proliferate, the fewer facts we preserve and the further we drive the thylacine into oblivion. Such concerns loom large over the evidence for the number of thylacine captures in the 1920s and 1930s proposed in the current literature and it is essential that this key indicator of the species' distribution in the years immediately prior to it becoming unobtainable is understood as accurately as possible. Consequently, any defensible assessment of the thylacine must be evidence-based and evaluating the quality of that evidence and identifying its potential shortcomings is an essential first stage.

No contemporary primary evidence of a thylacine's capture has ever been recovered, except in the rare instances when the captors advertised their animals for sale or display. From these advertisements, we would know that a thylacine had been captured around the date that they were placed, the name of the captor and, potentially, a general location, but little else. Only with contemporary primary sources from the animal's purchasers, or additional primary or secondary information supplied decades later, can these captures be better understood.

There is no more objective source of evidence than the records of sale. People may later mistake one decade for another, the identity or location of captor and capture, even the species they have entrapped, but money tends to be meticulously noted right down to the ha'penny. With the limited number of buyers by the 1920s, the records in question are restricted to the minutes of Hobart City Council's Reserves Committee (HCCRC), the body responsible for the purchase of Beaumaris Zoo's stock while under the control of Hobart's City Council, and the notebook of James Harrison. Yet, even here things are not as clear as it might be hoped. Though comprehensive, the HCCRC minutes rarely directly refer to the thylacine and specifically mentions purchases on only two occasions, the Loveluck, and Murray animals, in 1923, and 1925 respectively. The remainder are listed under the name of payee and amount only, along with almost every other service and product purchased on behalf of the Council. Consequently, interpreting the HCCRC's payments requires detailed further research. Currently, opinion as to how many thylacines they record varies between nine and eleven, equating to six adults and five cubs (Linnard *et al.* 2020), about a dozen (Sharland 1962) to in excess of 28 (Paddle 2012). Although limited to only four pages that contain references to the thylacine, Harrison's notebook confirms that it was he, not Beaumaris, who was the primary figure in the thylacine trade during the 1920s. However, those four pages, which probably relate to around 1928, provide little insight into how many thylacines Harrison traded over the course of his two decades dealing in the species.

The most detailed testimony is usually recovered decades after the event, yet formal published consideration of the testimonies of thylacine witnesses or captors is rare within the first four decades after 1936. Generally, they are recorded in popular sources, which requires an awareness of journalistic license, or else in unpublished notes, which are frequently represented in the literature in succinct form, often in the context of the secondary author's interpretation. Examination of the writings of Michael Sharland between 1937 and 1962 illustrates the extent and speed that even the most respected and authoritative sources can become confused (see pp. 104–6).

Even retrospective *primary* testimony is not pristine; predictably, as time elapses errors often arise. The Delphin captures were reported in July and August 1930 (*Advocate* 1930a, 1930b), yet five decades after these articles were penned, and without any further reference points to the events, by the 1980s the Delphin family dated them to 1933 (D. Enkler, *pers. comm.* 29 May 2020). Tyenna resident Algernon Chaplin (1954), undoubtedly acquainted with Elias Churchill, stated: '20 years ago one was captured in the fields (Mt Fields) by Churchill'. However, all the evidence supports Churchill's capture being made in 1923, not 1934, as Chaplin implies, suggesting that in three decades, his dating had veered off course by 11 years. The Delphins were doubtlessly influenced by the popular proliferation of the idea of the last capture taking place in 1933. Chaplin's error is less explicable, but warns us that similar mistakes may exist in other testimony and without the benefit of a transaction recorded in a contemporary source, are potentially undetectable. Consequently, when considering the dates of retrospective accounts of wild sightings, a high degree of caution is essential.

Many of the secondary witness accounts, although inherently well meant, and undoubtedly honest, are exemplified by a letter written by Kathleen Jenkins (1981). In addition to attributing the Mullins capture to Churchill, and vice versa, Jenkins makes reference to bushman Alex Nickols [sic]. Known, by the local children as 'Old Black Alex', who 'supposedly' led a leashed thylacine along the main road from Fitzgerald to National Park, where he, 'kept it tied up like a dog for some time'. Jenkins acknowledges: 'my father talked about it but I didn't see it'. No other reference to these incidents have ever been recovered, which is concerning in itself.

Nickols, actually Nicholls, though in fact this was an alias, the true name of the Syrian-born bushman being Melekh, or possibly Melike, Angoola, definitely existed (Fig. 16). However, there seems to be no reference to his thylacine-taming exploits beyond Jenkins' letter, so how seriously should we take it? Objectively, as we have no further evidence it is impossible to say. Subjectively, it seems unlikely. If it were true the question arises, why did Angoola, who was arrested for vagrancy in 1928, fail to capitalise on his thylacine's worth by selling it. Yet, it has been cited as evidence of the thylacine's tractability (Paddle 2000).

Crucially, numbers appear to become inflated with an over-reliance on secondary testimony. Alison Reid recalled that her father retrieved two thylacines from the south-west. It was believed that this was confirmed by two entries in the HCCRC minutes (Paddle 2012). On 27 March (1928a), the HCCRC minutes note that Reid was to be sent to Port Davey to acquire thylacines. On 17 April 1928, a payment of £20 was made to Reid (HCCRC 1928b). Yet, Reid never went to Port Davey; instead, he was sent to a conference in Melbourne (*The Mercury* 1928b), the £20 being paid for his travel expenses, removing two thylacines from the historical record.

Fig. 16. Melekh Angoola alias Alex Nicholls, 1928. Tasmanian Archives: Gaol (Branch) Prisoners Record Books, GD63-1-6 c1928 p. 701.

Other secondary accounts are presented with the teller's acknowledgement of their testimony's limitation. Ernest Bond related, but refused to confirm, an account he had heard of an alleged capture by Elias Churchill at Adamsfield in 1926, though Smith (1981) repeats Bond's account without acknowledging the uncertainty.

Significant secondary evidence is also preserved in the state's media. Thylacine captures were regularly, but not always, reported in the newspapers, so we cannot rely on them as an absolute guide. When captures were reported, the articles weren't especially prominent nor were they syndicated among all of the state's newspapers. A capture at Waratah for example might be reported in Burnie's *Advocate*, but not in Hobart's *The Mercury*, nor even in Launceston's *Examiner*. Although this is not an issue today with the overview, and keyword search function, of Libraries Australia's Trove, previously it has definitely led to confusion as people tended to be most aware of, and to over emphasise, events local to them. Consequently, many areas have at least one claimant for the captor of the 'last' thylacine.

In general, the Tasmanian press reports were of varying accuracy. When contemporaneously written by interested parties such as Sharland, they give concise information that appears accurate when examined against other sources, such as the HCCRC minutes. That said, the majority of the newspaper articles were anonymously written and in some cases, at least to an extent, are demonstrably inaccurate. Then as now, the media were not restricted to the written word. We have no idea what information was broadcast on radio during the 1920s and 1930s, only that some was, and all of it is now lost. Beyond speculation, we cannot judge how accurate this information was or how influential it was on witnesses relating their testimony years or decades later.

An elephant for the kiddies: the trade in live thylacines

Gareth Linnard

Arthur Robert Reid, Curator of Hobart City Council's Beaumaris Zoo (1923–37), believed the key to attracting paying visitors was displaying foreign species. To a large extent the evidence suggests he was right, although given the cost of maintaining such animals and the size of Hobart's population, his model was unsustainable. During the zoos formative years, Hobart's council were in full agreement with Reid and aided him in placing as many exotic animals in the zoo on the Domain as possible. Essentially the unwanted excess stock of larger mainland and international collections, these animals were rarely purchased; instead, they were acquired by exchange for Tasmanian fauna, which the rest of the world considered to be exotic. Wallabies, quolls and devils are of course, cheaper to feed than lions, leopards and tigers. By the time the implications of this were realised, the Zoo had become a municipal albatross around the city's neck and among the city's Reserves Committee, confidence in Reid's stewardship was faltering.

Of all Tasmania's indigenous fauna, the overseas zoos were most interested in acquiring devils and thylacines (Fig. 17), although exchange was by no means the Beaumaris Zoo's

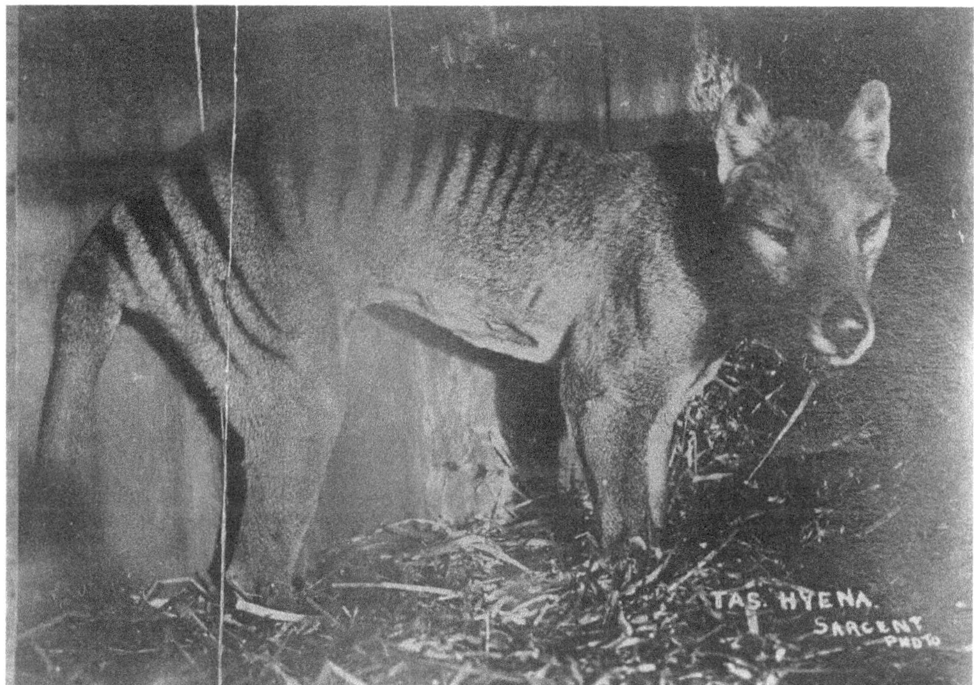

Fig. 17. Tasmanian tiger [on reverse – Photo by J. A. Fletcher of a Tasmanian Tiger Mrs Roberts had], personal papers of Nan Chauncy, Tasmanian Archives: NS351/1/26.

only motivation in acquiring these species. Both added a considerable scientific cachet to Hobart's otherwise unremarkable zoo and like most educated people, and a great many with less opportunity to become so, Reid had a personal interest in Tasmania's indigenous carnivores (*Illustrated Tasmanian Mail* 1934; Mitchel and Sargent 1996). Yet Beaumaris had a competitor, equally determined to secure thylacines: James Harrison, based in the north-western town of Wynyard. As the 1920s progressed, their relative success in securing thylacines provides compelling evidence of the contraction in the species' distribution.

Like Mary Robert's private Beaumaris Zoo, the Council's Domain Zoo intended to trade thylacines, as the inclusion of a £40 resale price in their 1923 price list demonstrates (HCCRC 1923a). However, in July 1925, their plans were scuppered by an elephant, for which they had arranged to exchange a pair of thylacines with London animal dealer George B. Chapman (HCCRC 1925a).

Led by the state's formidable Police Commissioner, Colonel John Ernest Cecil Lord (Fig. 18), the State Advisory Board, responsible for administering the *Animal and Birds Protection Act* 1911, was not concerned with attracting visitors to the zoo by securing 'a baby elephant for the kiddies' (*The Mercury* 1924a). The Board members were, however, concerned with preserving the thylacine in its home state, or at least some of them were. So, while Beaumaris was reluctantly given permission by the Board to export two thylacines to London in 1925, provided both were female, it was hinted at, very heavily, that no further such accommodation would be granted (Lord unpublished a). Consequently, just 2 years

Fig. 18. Colonel John Ernest Cecil Lord, a forgotten but important advocate for conservation of Tasmania's wildlife. Courtesy: Tasmania Police Museum Collection.

5 months after opening, Beaumaris Zoowas no longer a dealer, but strictly a purchaser of thylacines.

Harrison, on the other hand, was purely a dealer. His main customer was orthopaedic surgeon Sir Colin MacKenzie of the Australian Institute of Anatomy, who believed thylacine anatomy held clues to the treatment of human musculoskeletal ailments. He also had very deep pockets; all individual thylacine payments by Beaumaris were for half of what Harrison (unpublished) recorded MacKenzie as paying. Consequently, with access to a more lucrative interstate market, despite their otherwise cordial and cooperative relationship, Harrison had no motivation to sell thylacines to Beaumaris, leaving the zoo to secure its own. The sole exception being a female with two short-lived dependant cubs in August 1924, purchased for £46/10/0, a price roughly equitable to, if slightly lower than, MacKenzie's rate for a single thylacine (*The Mercury* 1924b; HCCRC 1924a). Despite the imbalance in resale price, there is evidence Beaumaris purchased thylacines for a higher sum than Harrison. Jordan (1987) remembered receiving £20 from Harrison in 1928 or 1929 for his thylacine, while Arthur Murray and Roy Delphin each received £25 from Beaumaris in 1925 and 1930 respectively (HCCRC 1925b, 1930). Jordan aside, the very fact Beaumaris secured thylacines from Murray and Delphin at all, both of whom were based in Waratah in the north-west and within Harrison's sphere of operation, suggests that they could compete with him on price. So why is there no contemporary evidence of them acquiring any further specimens?

The Murray and Delphin captures share a commonality not seen in any other northern capture of the era; both were advertised in the press. By the Murrays in 1925, in Launceston's *Examiner* and significantly, on three occasions in *The Mercury*, Hobart's main newspaper (Murray 1925). The Delphin's displayed one of their thylacines at the Burnie Show and advertised the forthcoming attraction in the town's newspaper, the *Advocate* in 1930. Undoubtedly, this is how Beaumaris became aware of both thylacines.

Harrison's purchases, which were only documented as being made in the north, were always reported retrospectively by the press, suggesting he was securing thylacines through local connections, as Beaumaris had been able to do in the south-west in 1923 when they successfully acquired all the known captures from the region. Yet, despite their public requests for thylacines and advertisements in *The Mercury* (1925; Brain 1928a) and *Examiner* (Brain 1928b), after 1923 Beaumaris failed to secure any more, except of course the two from the north-west.

By contrast, Harrison's notebook shows he had procured at least four thylacines around 1928 alone. On 15 May 1929, *The Mercury* (1929) reported an assessment of the situation by the director of the Tasmanian Museum and Art Gallery, Clive Lord. Inevitably referring to Harrison, Lord stated of the thylacine that: 'one man had secured a corner in them and neither the zoo nor the museum had been able to get a specimen during the past three years'.

The surviving evidence is sparse, yet it does not support Harrison's primacy being based on his offering a higher purchase price than Beaumaris, though he clearly could have adopted this approach. Not in doubt is that Beaumaris began to seek specimens from the north, immediately after the last definite contemporary evidence of the species being recovered alive from the south of, what is now, the Lyell Highway. With the incentive to any south-western captor to present a live thylacine to Beaumaris, created by the prices the zoo was offering, this is unlikely to be a non-causal relationship.

A surprising number of skins: the trade in dead thylacines

Gareth Linnard

Although the trade in live specimens suggests the thylacine was unobtainable from the south-west from the mid-1920s onward, visiting naturalist Sydney Jackson (1937) commented in 1928 that he was surprised at the number of thylacine skins on sale in Hobart and Launceston. We cannot know how easy Jackson was to surprise, nor the age of the skins on sale, but he appears to have been implying that they were readily available. If so, as far as Hobart is concerned, this is at odds with a 1929 statement by Clive Lord (Fig. 19), 'neither the zoo nor the museum had been able to get a specimen during the past three years' (Lord unpublished b), although at £10 per body; there was a strong incentive to supply them. As for their availability in Launceston, the city's museum, the Queen Victoria Museum and Art Gallery, received no recently killed specimens after 1912 (S. Sleightholme, *pers. comm.*).

Further disputing Jackson's statement is a letter from May 1925 by Professor Frederick Wood-Jones (unpublished) of the University of Adelaide; writing to Oldfield Thomas of

Fig. 19. Clive Errol Lord, Director of the Tasmanian Museum and campaigner for the thylacine's preservation. Pretyman family, photographs and glass plate negatives collected by ER Pretyman (NS1013). Tasmanian Archives: NS1013-1-1841.

London's Natural History Museum, he stated, 'skins seem quite impossible to obtain: & I only secured two imperfect skulls'. Although the disparity between Jackson's and Wood-Jones' experiences could be explained by a sudden influx of skins around 1928, Lord's 1929 comment makes this seem unlikely. We do have evidence of a number of old skins being in private hands at this time and from the *Advocate* (1927) one presumably fresh skin was in a Burnie furrier's possession. Yet, with the exception of a body of a thylacine trapped near the Florentine River by Paddy Hartnett, acquired by the Tasmanian Museum and Art Gallery in 1926 (*The Mercury* 1926), the otherwise total absence of evidence for thylacine skins coming into Hobart is just another rather negative indicator of the species' persistence in the south-west.

A further unknown variable is the private trade in skins, with overseas buyers alluded to by Lord and Butler (*Examiner* 1922), Stead, President of the Wildlife Preservation Society of Australia (*Advocate* 1930c) and Guiler (1986). The export of thylacine material from Australia was banned in 1930, initially suggesting there was an issue significant enough to be legislated against. Despite this, evidence for a significant private export market of thylacine specimens during the 1920s and 1930s is absent, as confirmed by Dr Sleightholme, Director of the International Thylacine Specimen Database. Excluding the undocumented animals acquired from Harrison by McKenzie, which are now held by the National Museum of Australia (Canberra), of mainland collections, Sleightholme states: 'With the exception of the

NMA Selby-Wilson skin, all of the specimens in mainland Australian museum and university collections acquired after 1920 were derived from zoo stock' (S Sleightholme, *pers. comm.*).

As for American museums, Sleightholme states: 'All specimens in American collections are known, but only two are noted as being obtained in the 1920s that were not derived from zoo stock. These comprise a skin and a skull and skeleton held in the Smithsonian's collection' (S. Sleightholme, *pers. comm.*). Both of the Smithsonian specimens were collected by Charles Macaulay Hoy in 1921. European collections hold one definite wild kill from the era, a skull in Glasgow's Hunterian Museum, acquired in 1925 from Professor Theodore Flynn, and taxidermy in Frankfurt's Senckenberg Research Institute, sold by Hamburg dealer J. F. G. Umlauff in 1924, though it is not clear when it was acquired by the latter (S Sleightholme, *pers. comm.*).

Although we do have advertisements from the 1920s of thylacine skins being auctioned from house clearances in Tasmania, there are none after 1940, indicating they probably did become in short supply some decades earlier. Beyond the donation of a skin to the Tasmanian Museum in 1946, a skin purchased from New Zealand by the National Museum of Australia in 2018 and a Sydney auction that included a thylacine skin (Shapiro 2008), there is little evidence of privately held thylacine material. In fact, the International Thylacine Specimen Database notes only four skins that are currently held in private collections.

It therefore seems, as implied by Wood-Jones, the evidence for privately purchased specimens, academic or otherwise, cannot account for a significant number of potential kills. Possibly the best interpretation of the comments of Lord, Butler and Stead is that the thylacine may have been periodically singled out when discussing the export of Australian fauna due to its scientific importance, rather its frequency of export or, perhaps more plausibly, as a result of the frustration among Tasmanian institutions born of Harrison and MacKenzie's dominance in the trade in live and dead thylacines. Leaving us with what may be either a large gap in the evidence or further evidence of the species' terminal decline. In the former case, it seems that we should start by noting that it is very curious no contemporary mention of a private trade was ever made beyond the one *Advocate* advertisement. Then ask why, if the skins were so ubiquitous around Hobart and Launceston as Jackson suggests, the species was declared endangered and even extinct periodically in the state's newspapers and why live and dead thylacines were so relatively expensive?

1920s and 1930s: the road to 1936

Gareth Linnard

This chapter must necessarily open with a caveat to the effect that it is certainly not suggested to be a definitive catalogue of the thylacines killed or captured during the 1920s and 1930s. It is a collation of all currently known contemporary reports or those supported by additional evidence. It is undoubtedly an underestimation.

The earliest thylacine kill of the decade was reported on 8 June in the *Examiner* (1920), from the north-east, and would be the final report from the eastern half of the state. The following year, a kill was reported at Russell Falls bordering National Park in the south-west (*Hobart Town Gazette* 1921). The Tasmanian Museum paid £5 for a skin in March 1922 (*The Mercury* 1922b). Unfortunately, this skin cannot be traced among the current records nor was the capture location reported. In the following July, the *Advocate* (1923) reported James Harrison showing an unsexed thylacine of unspecified age to a party of visitors. Also in 1923, six individuals were captured alive in the south-west: Churchill's male (Linnard *et al.* 2020), Loveluck's female and Walter Mullins' group of a female and three cubs (HCCRC 1923b, 1924b). Clem Penney killed a female near Waratah in 1924 (Haygarth 2016), the kill being recorded for posterity by a photograph of the young Penney proudly holding the body of the dead thylacine (Fig. 20).

The final contemporary reference to a family group, a female with two pouched young, was sold to Beaumaris Zoo by Harrison in August 1924; however, there is no evidence to support that the young survived long enough to be placed on display (HCCRC 1924a; *The Mercury* 1924b). In May 1925, the Murray brothers of Waratah captured a juvenile female, which they sold to the Beaumaris Zoo in August 1925 for £25 (HCCRC 1925b). On 4 August 1926, Tom Harman and Alf Forward reportedly snared two individuals, one

Fig. 20. Clem Penney holding his trophy. Tasmanian Archives: Tasmanian tiger (thylacine) – killed by Clem Penney; Miscellaneous Collections of Photographs PH30/1/6303 c1920.

escaped, the other, a female, choked to death in the snare (*Circular Head Chronicle* 1926a). Its body was subsequently sold to Harrison. On 29 September 1926, the same paper again reported a further encounter between Forward and a thylacine. Trapped in his snare, the animal bit him on the hand and escaped (*Circular Head Chronicle* 1926b). The *Mercury* (1926) reported that a thylacine, captured near the Florentine River by Paddy Hartnett, was mounted for the Museum by Alison Reid in October 1926, where it still stands today on public display, this represents the last contemporaneously recorded thylacine killed or captured in the state's south. The *Advocate* (1927) reported a skin on display at the Ulverstone Show.

1928 saw the capture of a juvenile thylacine, reportedly on 17 March, at the Arthur River in the north-west by Messrs Chester and Hope (*Advocate* 1928). Harrison's notebook (Harrison unpublished) also records an order for three thylacines by Perry Brother's Circus, the undated sale of one of which is recorded. Although no further details are recorded in Harrison's notebook, on 17 December, Sydney's *The Daily Telegraph* (1929) reported the circus was in possession of 'two Tasmanian tigers'. A further four sales are recorded in Harrison's notebook, two of which are dated to 11 May 1928 and listed to MacKenzie. Harrison also records the purchase of two thylacines, likely accounted for among the sales, and what appear to be two thylacine skins.

Adye Jordan (1987) dated his and brother-in-law Frank Gleeson's capture of an unsexed juvenile, also on the Arthur River, to both the same year as another 'small' one with a broken leg that was captured in the same area (see Chester/Hope capture above) and to 1929. They also recalled a further two thylacines captured the same night, both of which were choked by their snares.

Only three thylacines are known from contemporary records from 1930. On 13 May 1930, Wilf Batty of Mawbanna is known to have made the last fully documented kill of a thylacine. The Batty kill, a male, is well represented in the photographic record. On 10 July 1930, the *Advocate* (1930a) reported a capture by the Delphin brothers, though it was actually Daniel Delphin and his nephew Gordon; this animal, a male, was sold to Beaumaris in October 1930. A second capture by the Delphins, this time by twins Roy and Gordon, was reported on 11 August (*Advocate* 1930b); the female died in the family's care. This was the last contemporaneously recorded contact with the thylacine in the wild.

Another individual thylacine was reputedly killed by a snare on the Pieman River in 1930 by District Surveyor Charles Selby-Wilson of Zeehan, although it is also sometimes said to have been shot. The latter fate seems less likely, as the skin, which includes the head, shows no obvious signs of a bullet or shot wound. The provenance of this skin appears to be based on a note supplied by the family who kindly donated it to the Australian Museum stating it was captured in 1930 or 1932. The author is not yet satisfied that this adequately supports the skin as being from the 1930s and as such, it will be excluded as a definite confirmed wild kill.

Given the mechanism of capture thylacines were subjected to (Haygarth 2017), a higher number of live captures than kills seems unlikely, suggesting that a number of the latter went unreported.

About 1935: the Churchill capture

Gareth Linnard

The most prominent position in the literature, still persistently repeated in many popular articles, is that the last known thylacine was captured in the Florentine Valley, south-west Tasmania, by Elias George 'Churchie' Churchill (1892–1974) in mid-1933 (Smith 1981; Paddle 2000). Each of the four known confirmed kills or captures in the 1930s were made in the state's north and although the evidence supports that Paddy Hartnett killed a thylacine near the Florentine River in 1926 (N. Haygarth, *pers. comm.*), no other definite evidence can be recovered for the species in its former strongholds in the state's south after 1923. If proven correct, the Churchill capture would significantly expand the range of the species in the 1930s, both geographically and temporally.

An article appearing in *The Australasian* dated 20 January 1934 is the only contemporary account of Beaumaris' acquisition of its final thylacine. Describing his visit to Beaumaris Zoo in mid-December 1933, zoologist David Fleay (1934) wrote: 'First and foremost is a fine male Thylacine, in fact the sole member of its kind in captivity today. It has thriven in Hobart for the last three years.' A comparison between the images taken by Fleay on his visit in December 1933 and by the photographer Ben Sheppard in May 1936 confirms beyond any doubt that this individual was the last captive specimen (Sleightholme *et al.* 2020).

At the time of his visit to the zoo, Fleay was in Tasmania to interview for the position of Director of the Tasmanian Museum, according to his daughter, motivated by his interest in preserving the thylacine (Fleay-Thompson 2007). It seems likely he would have carefully evaluated any information provided to him on the species. In light of the immediacy of Fleay's account, and that he spent his visit in the company of the zoo's curator Arthur Reid (Fleay 1934), his implied date of arrival for the last captive specimen, around late 1930, is extremely persuasive. Accepting the Churchill scenario compels us to accept Fleay made an error of around 3 years.

Fleay's *Australasian* article of 20 January 1934 was written no more than a month after his visit to Beaumaris. Yet, a review of the literature suggests that rather than forming the basis of opinion on the last captive's origins, the truths within were forgotten. What replaced it in the following years was an alternative view, complete with the capture dated to 1933, beginning, in lasting publications at least, with *Tasmanian Wildlife*, by prolific Tasmanian author, journalist and naturalist Michael Sharland (1962). Although Sharland did not name Churchill as the captor, he did locate the capture in the area he was known to have trapped during the 1920s, the Florentine Valley. Reflecting this belief, Paddle (2000) cites a letter by Sharland written in 1972, stating that he did not have, 'exact knowledge of where the last Thylacine to be kept in the old Hobart Zoo was trapped. The only record I have is that given to me by the then Zoo Manager, A. R. Reid, that this particular animal came from the Florentine Valley'. That claim, though, is directly refuted by the surviving documentary evidence (Linnard *et al.* 2020; Sleightholme *et al.* 2020).

Paddle (2000) also suggests Sharland believed Churchill was the captor and includes a further excerpt from his 1972 letter: 'you may also be interested in an interview with Elias

Churchill' (Sharland 1972). Yet, the entire cited content of the letter is reproduced and had Sharland explicitly named Churchill as the captor, it seems a curious omission on Paddle's part not to directly quote that statement. Without any context, the portion quoted is ambiguous and it is unclear whether Sharland was simply referring to the 1957 interview he carried out with Churchill. The interview, which was conducted in preparation for an article, one of a series in that year motivated by interest in the species stimulated by a Fauna Board expedition, discussed Churchill's general experience with the thylacine, not in connection with a claim that he was the captor of the last known example. Had Sharland believed Churchill captured Beaumaris Zoo's last thylacine in 1972, this is contrary to the statement he published in his 1957 article based on Churchill's primary testimony; that Churchill was the captor of not the last, but the first of Beaumaris' thylacines (Sharland 1957). As the author of the article that reported this animal's arrival (*The Mercury* 1923), Sharland would have known when the first thylacine on the Domain was supplied (Sharland 1923). Sharland appears to have confused Churchill having played a *part* in the Mullins capture, with him being the captor, though as a family group of a female and three cubs captured in 1923 none would be candidates for the last captive. However, in an interview given to *The Mercury* (1964), Churchill described only *one* live capture. Consequently, there can be little doubt that after listening to his testimony in 1957, Sharland could not have considered Churchill as the captor of the last thylacine. Quite who put his name to the last capture is unclear, but first recovered in Smith (1981), it certainly wasn't Churchill himself nor is there any direct evidence that it was Sharland.

Although it was believed that between 1932 and 1934 the minutes of the Hobart City Council Reserves Committee, and with them any possible record of a 1933 purchase, had been lost (Paddle 2000), in fact the committee itself had been temporarily disbanded and during 1933 responsibility for Beaumaris' finances was transferred to the Council's Health Committee, whose minutes are preserved. On examination, the Health Committee minutes reveal no reference to a thylacine purchase.

With Beaumaris' final recorded thylacine payment on 30 October 1930 to Roy Delphin for an animal captured in the north, and with no record in the Health Committee minutes of a purchase in 1933, is there any evidence to corroborate Sharland's claim the capture took place in 1933? A series of articles, beginning on 10 February in *The Mercury* (1937a), reporting the Tasmanian Fauna Board's response to a letter discussing the thylacine by Cecil Ryan to the State Premier Albert Ogilvie, suggest not. Implying an entry date of 1929, Ryan's letter incorrectly stated that the last captive had been in Hobart for 8 years, an error Sharland would repeat in *The Mercury* just 10 days later (Peregrine 1937). Ryan's letter was prompted by a conversation with his cousin, Albert S Le Souëf (Director of Taronga Park Zoo), who like Sharland was a member of the Royal Zoological Society of New South Wales. So it's uncertain whether Sharland copied the error or was its originator, passing it on to Ryan via Le Souëf.

It is clearly apparent that Sharland, resident in Sydney during the 1930s, had no knowledge in 1937 of when Beaumaris Zoo received its last thylacine. Although this raises significant concerns over Sharland's date and, significantly, excludes him as a primary witness who preserved a continuous knowledge of the last thylacine's capture, a secondary account would still be valuable if evidentially supported. Yet, even in this regard, Sharland's own writings seem to undermine him. A paper trail of articles by Sharland, Fleay and the naturalist Alec Chisholm, himself a visitor to Beaumaris in 1933, creates the definite impression that Sharland's change from 1928 to 1933 resulted from a shared misunderstanding among them.

The confusion, which developed over the following three decades, seems to stem not from the date of the last thylacine's entry, but of its death. The HCCRC (1936a) minutes confirm that the last captive survived until 7 September 1936and all three men reported the death within 6 months of its occurrence, describing it as occurring recently (Chisholm 1936; Fleay 1937; Peregrine 1937). Yet, in the years afterward, as their recollections faded, Fleay, Sharland and Chisholm, who certainly knew each other, converged on an erroneous date of 1933 for the animal's death. Sharland (Peregrine 1939) reported it as 'about 1935' and Fleay (1940) seems to have become confused, making the ambiguous statement that the thylacine had died 'not so many months after my visit', implying early 1934.

Chisholm (1945) became to the first to give the year of death as 1933 and 5 years later, Fleay was quoted as doing the same (*The Argus* 1950). Not to be outdone, Sharland (1962) upped the ante, dating both the death and capture to 1933. If any one thing should have alerted subsequent researchers to problems with Sharland's date, it was his claim that Beaumaris held its last thylacine for less than a year, but it did not. With no further supporting evidence, Sharland suddenly dating the capture to 1933 begins to look suspiciously like a typo, and one that has wasted a great deal of time. Yet the confusion and misinformation surrounding the Churchill capture is an important lesson to apply to wider historical thylacine research. Records are scattered and incomplete, and it is easy to incorrectly re-assemble what remains.

Two little tigers: the Delphin capture

Gareth Linnard, Mike Williams and Branden Holmes

> There has been so much written about the Tasmanian Tiger that I would like to tell my story or rather my twin brothers Gordon, Roy Delphin story & Uncle Dan Delphin late of Waratah Tasmania.
>
> – Dorothy Gould (no date)

In 1930, Burnie's *Advocate* (1930a, 1930b) reported that the Delphin family had captured two live thylacines. The first report on 10 July stated: 'Messrs D. and A. Delphin were fortunate to snare a Hyena on Monday evening'. And on 11 August: 'Some time ago Messrs. Delphin Bros., when snaring captured a hyena in the vicinity of Mt. Bischoff. Last week they were successful in capturing a smaller specimen of the same species'. The first animal was captured by 59-year-old Dan Delphin (Fig. 21) and his 19-year-old nephew Roy, the second, by Roy and his twin brother Gordon.

Further details of one of the captures, most likely the first, are provided by Godfrey (1984): 'It was caught on a "treadle" snare on the West Coast in thick, overgrown country and brought back to Waratah carried in a chaff bag slung on a pole'. Both animals were

Fig. 21. Dan Delphin (left) with brothers John (centre) and Augustine (right). Courtesy: Delphine Enkler.

taken back to the Delphin house, which stood on Hall Street, home of Mary Delphin, sister-in-law to Dan, widow of his late brother Augustine and mother of Roy, Gordon (Fig. 22) and 11 other children, including Dorothy Delphin, later Gould, who celebrated her 7th birthday on 3 September 1930 while at least one of the thylacines were in the family's possession. Both thylacines were placed in a wooden shed in the garden, with a fenced-off earthen-floored pen attached, which usually doubled as a chicken coop or Gordon's pigeon loft (Fig. 23).

The Delphins are adamant the female died while in the family's care.In an undated letter written sometime during the 1980s, Dorothy Gould wrote: 'Gordon and Roy caught a second one, she died shortly after'(Gould no date). We don't know exactly when her death occurred, except that it was at some point prior to 24 September when the Delphins began to advertise in the *Advocate* (Delphin 1930) that a single thylacine would be available to view at the forthcoming Burnie Show.

The male would go on to be sold to the Beaumaris Zoo, for which a payment of £25 was authorised to 'R. Delphin' on 30 October 1930 (HCCRC 1930). Payment noted the thylacine would have arrived at the zoo some weeks earlier, its purchase having already been reported in *TheMercury* 13 days earlier as having been made the previous week (*The Mercury* 1930a). It seems highly likely that the sale was negotiated at, or immediately after, the Burnie Show, which ended on 2 October.

Fig. 22. Gordon Delphin. Courtesy: Delphine Enkler.

The female's time spent in the family's care could not have exceeded 48 days; the male's could not have exceeded 102 days. Denoting the final occasion when the world's last confirmed thylacine saw one of his own kind, the young male and female's time together in the pigeon loft overlapped by a maximum of 48 days. The Delphins found the temperament of the two young thylacines very tractable: 'He tamed down a lot after a while. He'd growl and jump about but he never offered to bite. Mum used to feed him warm meat in a little billy can. He always liked his warm meat' (Godfrey 1984). A further gauge of the Delphins' experience of the animals' behaviour was that 6- to 7-year-old Dorothy was allowed to feed milk to the thylacines. If the family found the thylacines easy to live with, their dogs certainly did not. Dorothy recalled: 'When my brothers let them (thylacines) into the pen at times, their dogs were so frightened of the tigers that they wouldn't come out of their kennels until the tigers were back in the shed' (Gould no date).

Several witnesses remembered the thylacines at Gordon's pigeon loft. Ian Rist, son of Alfred and Maisie Rist of Waratah, recalled: 'My dad remembered the two young Tigers the Delphin Bros. had at Waratah in an old rebuilt chook pen. My mother actually had some photos of the two' (I Rist, *pers. comm.*). Sadly, the Rist photographs are now lost, but a more detailed description survives from Margaret Breaden, who recalled coming from Devonport to see the thylacines: 'I think it was while I was at Waratah this happened. There were two

Fig. 23. Mary Delphin in front of the thylacines' temporary home. Courtesy: Delphine Enkler.

young men, Delphins their names were. I can't remember their Christian names. And they had caught these two, they weren't full-grown tigers. And they had them in a tent, I remember it vividly. With dog collars round their necks, and chains. And they spoke of taking them around the shows and showing them at the sideshows. But after that I can't remember what happened to them. I can remember seeing the two little tigers' (Wisbey 1994).

It's notable that Breaden describes the thylacines in a tent, rather than in the pigeon loft, possibly indicating that she saw them at one of the country shows. Aside from this possible reference, no primary source has been recovered describing either animal at a public show; however, Godfrey (1984) does provide a secondary reference: 'Those who usually did the showing were his (Gordon's) uncle Dan and sister Pearl with Charlie Fulford and his daughter. In the act around the showing, Charlie Fulford played a melodeon and the girls danced'.

The Delphin capture satisfies all four identification criteria as noted by Fleay (1934). All the recovered witnesses support the Delphin's assertion that both thylacines were sub-adults. October is within 2 months of the date of arrival given by Fleay (1934) as 'three years' before his visit in December 1933. Roy identified the scar on a photograph of the last captive as matching the injury to the thylacine he sold to Beaumaris Zoo and Dorothy wrote: 'the thylacine that ended up in the Hobart Zoo, was easy recognised by the snare mark on its hind leg (treadle snare)' (Gould no date).

Sex was the singular stumbling block in identifying the Delphin animal as the last captive due to the wording of an article published in *The Mercury* on 17 October 1930: 'In place of the last one which died 18 months ago from kidney trouble, a Tasmanian wolf was purchased for the Beaumaris Zoo last week. She was caught by a trapper at Waratah together with another which died from wounds, and as far as is known, is the first of its species to be taken in Tasmania for the past four years. The skin of these animals is very tender, and it is seldom that the flesh heals without difficulty, this, no doubt, causing the death of the other wolf with whom this **female** specimen was trapped.' (*The Mercury* 1930a). There can be little doubt that this article referred to the Delphin animal, so the statement that it was female initially appears to preclude it as the last captive. Yet the Delphin's testimony is at odds with the anonymous journalist.

Roy Delphin referred to the surviving animal as 'he' (Godfrey 1984) and in the draft letter written by Dorothy, she consistently describes the animal captured in August as 'she', but to the July animal as 'it', suggesting the sexes were differentiated. This differentiation is supported by Delphine Enkler: 'I assumed the first one was male no-one ever said it was a female, only that second one was a female' (D Enkler, *pers. comm.*). The author suggests that the most likely explanation is that *The Mercury* journalist, who misquoted the date of the previous thylacine capture by 2 years (*Advocate* 1928), was poorly informed. Lastly, the subject of the Doyle film, taken on 23 December 1930, is male, the only candidate for this animal being the Delphin thylacine.

The fate of Delphin's first live capture is currently contested, with Linnard *et al.* (2020) advocating that it was the last captive. However, this argument is countered by Paddle (2012), who champions Elias Churchill as the captor of the last of Hobart's thylacines, and Sleightholme *et al.* (2020), who advocate James Kaine. Provenance disputes aside, the history of the Delphin family is known and is an important dimension to the story of Hobart's last thylacine.

Mary Delphin was determined that her younger children would have the opportunity to remain in school and gain a better education than her twin sons Roy and Gordon, both of whom had their education severely hampered by problems with their eyesight. Sadly, this was not to be. Despite Mary's determination, by the age of 11, Dorothy had left home to work. Although she always called Waratah 'home', in 1941 Dorothy moved to Melbourne where she spent the rest of her life and where her daughter Delphine was born (D. Enkler, *pers. comm.*). It is through Delphine's attention to her family's history and preservation of their correspondence that has provided the insight into the 'two little tigers' of 1930. Delphine met Gordon only once, shortly before his death in 1967; despite this, she remembers him very fondly. It was, however, her uncle Roy to whom she was particularly close, the co-captor of the last thylacine definitely recorded at Beaumaris Zoo.

Roy, convinced of the species' persistence, refused all interviews on the subject until his late seventies. When he did speak, he exaggerated the thylacine's ferocity, not to malign the species, but as he explained to his family, to dissuade anyone from disturbing them (D Enkler, *pers. comm.*). Dan Delphin died in 1934, 2 years before the thylacine he and his nephew had captured. Gordon Delphin remained on Hall Street until his death in 1967 aged 56, by which time the family had dispersed. Following his death, the Delphin house was seized in lieu of rates arrears and demolished, its exact location now lost. Roy Delphin spent most of his life in Tasmania, though between the late 1950s and 1965 he also lived in Melbourne.

When reading about the last known thylacines, it is easy to either forget their captors or to view them in simplistic terms biased by retrospect and modern attitudes. The Delphin's, Walter Mullins, and Arthur Murray, were neither cruel nor callous; this is evident in the treatment their captives received. In our gentler, more privileged time, it is easy for us to fail to consider the harshness of life in 1930s rural Tasmania and the financial relief a chance thylacine capture could offer. Thylacines were a significant, but unexpected source of income, and one that could not be ignored if the opportunity arose. In Roy's case, his experience with the two thylacines appears to have been more profound and never appears to have left him. Roy Delphin's headstone records his date of birth, death (which, much to his sister's annoyance, is incorrect) and mentions his relationship to those he loved, but also, embossed in the upper right corner, is a gold image of the thylacine he captured as a19-year-old, 59 years earlier.

The trail to the last Tasmanian tiger vision

John Doyle

On 28 July 1906, Harold Edgar Samuel Doyle was born at Aberdeen in New South Wales. As he grew up on the family farm 'Bundabulla', Muswellbrook, he developed a journalistic interest, producing his own family-orientated newspaper. The *Spring Creek News* was printed on butcher-style paper, which came with food deliveries to the mail box. Harold, or Ned as the family fondly called him, was the eldest of five brothers and a sister and he often used them in his paper in various roles, such as stock and station agents, auctioneers, insurance agents etc. His publication was comprehensive, including book reading, puzzle corner, poetry (often his own), wartime and peace treaty updates, rainfall, weather forecasts, gardening and much more. All were handwritten, published at least weekly and at the price of a halfpenny. He was 13 years old.

Although he continued his love of writing throughout his life, in time, another calling took over and he moved to Sydney to study at Moore Theological College. At completion of his studies, the now Reverend Harold Doyle married Mysie McCloy Colvin at St Paul's Chatswood, on 18 October 1930 and took up his first Parish at St Stephen's in Penguin, northern Tasmania, on 21 December 1930. By this time Harold had developed an interest in photography and had procured a Pathé Baby hand-cranked 9.5-mm movie camera and projector.

He had filmed Sydney Harbour scenes, including the building of the Sydney Harbour Bridge, and some parts of the newlywed's trip to Hobart on the ferry *Zealandia*. Throughout his time in Tasmania, Harold filmed such things as baby penguins, Mysie's pet possum, HMAS *Australia II* while visiting Burnie, trips to Gunn's Plains caves and Cradle Mountain.

In April 1934, Reverend Doyle resigned, due to the fact he was owed £71/4/8 salary and could no longer sustain a living. He and Mysie returned to the mainland where they held parishes in Gympie, Glen Innes, Emmaville, Barraba, Nowra and lastly Milson's Point, Sydney, before retiring. Harold continued filming throughout these years. When my father died on 1 May 1974, aged 67, I inherited his collection, which by that time included a Harmer and Heath 16-mm projector and many films that I had helped transport and set up at numerous showings in church halls for film nights.

In 1982, the National Film and Sound Archive (NFSA) announced a 'Nitrate Film Drive', to collect film that could either be copied to a more stable platform or stored in ideal environmental conditions to extend its life. Not knowing if some of my films were nitrate, I decided to have them collected and held by the NFSA for examination. Time moved on and so did we, to another home in a nearby suburb.

Memory of my films faded until I heard, or read, something that reminded me of them once more. It was October 2009, so I decided to make an enquiry on the outcome of their planned examinations. After initial phone enquiry and follow-up, between me and NFSA staff members, it was established that the NFSA had tried to contact me by mail regarding the films in July 1990, but it had been sent to my previous address. The correspondence had been returned to sender and at the time of my subsequent enquiry, I was advised that no record of my films existed.

My determined enquiries eventually put me in contact with a senior officer, who instigated a major investigation and search of unassessed material. Eventually I received news that most of my 9.5-mm film had been found and some had been digitised. I was pleased to be presented with them on DVD and decided to continue to leave the originals with the NFSA. I am still hopeful that the rest of my collection, one day, will be found.

In mid-March 2020, I received an email from Mike Williams, asking if he could please view footage held within my collection, titled 'Glimpses of Tasmanian Tiger in Zoo enclosure', as he was writing a book on the thylacine. Mike also asked if I could talk to his research colleague, Gareth Linnard, a leading archivist on the thylacine.

I was aware of the short footage of the Tasmanian tiger my father had taken, but had not spoken to anyone who had such knowledge and enthusiasm towards the animal, so I had no hesitation to allow them to view the film, answer their questions and help wherever I could.

Throughout the next few weeks our back-and-forth communications made it possible to narrow down the exact date that the film was taken (23 December 1930) and to come to the exciting conclusion that although it was only a few seconds of grainy vision, it was, in Mikes own words: 'Amazingly important historical footage! Gareth has been able to confirm that the film is the only footage discovered of a tiger cub and he says, "it's not just any cub either, it's the last cub". It is, without doubt, the most precious glimpse we've ever had of the tiger'.

I am so pleased that somehow chance, inquiry and perseverance have managed to bring the film from the possibility of total loss to the notice of people who genuinely care passionately about it. It has given me great pleasure to play a small part in something so important, not only to thylacine researchers such as Mike, Gareth and Branden, but now to my family as well, and to the scientific world. And thank you, dad.

The last known photograph of the Tasmanian tiger

Anthony Black and Gareth Linnard

One of, if not *the,* most recognisable images of the thylacine is that taken by Benjamin Sheppard at Hobart's Beaumaris Zoo, in its municipal incarnation on the Queen's Domain. Despite the proliferation of this image in popular and specialised sources, its origins have been unclear.

Fuller (2013) notes: 'Taken by Benjamin Sheppard on 24th January 1928. It is often said to be a portrait of the very last Thylacine but this is not the case'. An initial search of the history of the publication of this image suggests that in earlier instances neither attribution was particularly frequent, as it was usually presented as a generic illustration of the species, rather than being attributed to any date or individual. As is the case of the earliest recovered instance of its publication by Barrett (1944) and in two of the three recovered popular press articles of the 1950s in which it featured, those of Adelaide's July *Chronicle* (1949) and in January of Melbourne's *The Argus* (1957*)*. In each of these cases, except for a credit to Sheppard in Barrett (1944), no further information is offered. However, in an *Australasian Post* (1956) article the image is captioned, 'this photograph was taken when the last "tiger" was captured'. Griffiths (1972) captions the image as 'a pre-1930 thylacine', while in a booklet produced by the Tasmanian Museum and Gallery in 1985 it is stated that the image was taken at Battery Point, Beaumaris Zoo's location until 1923. The image appears in Eric Guiler's (1991) *The Tasmanian Tiger in Pictures*, but despite featuring on the title page in some versions and again within the main body, no information beyond 'author's collection' is offered, while Guiler and Godard (1998) credit the photograph only as 'private collection; Eric Guiler'.

Paddle (2012), probably the most influential author to discuss the Sheppard photograph, and certainly the one who gives it the most consideration, includes the image in his discussion of the effects of an epidemic disease he suggests the species was subject to during the late 1800s and early 1900s. Identifying Sheppard's subject as a sickly juvenile male, subject to this postulated disease, Paddle dates its entry into the zoo as the 24 January 1928. Furthermore, both Paddle (2012) and Fuller (2013) assert that the subject died the day after it was photographed. These assertions are referenced to an unpublished manuscript by the scientific illustrator Norman Laird, now held at Libraries Tasmania (NS1143) as part of a series of records of the species assembled by Laird.

Clarity was needed; fortunately, my request for information reached Libraries Tasmania archivist Anthony Black. We may forget just how dependent researchers are on the skills and tenacity of archival staff, if were not for Anthony's determined research, the true value of Sheppard's image (Fig. 24), and the historical information it confirms and refutes, would never have been realised.

In May 2018 we received an intriguing request from researcher, Gareth Linnard, a member of the Tasmanian Tiger Archives group. Gareth was interested in a file contained in the research papers of Norman Laird, purportedly relating to a thylacine housed at Beaumaris Zoo in Hobart around 1928. Gareth believed that the information in the file, as well as a

Fig. 24. Sheppard's notorious thylacine image, a far cry from the cub captured in the Doyle film. Tasmanian Archives: Tasmanian Tiger (Thylacine) at Beaumaris Zoo in May 1936. Photograph album, photographs, negatives, newspaper cuttings and miscellaneous papers belonging to Ben Sheppard (Photographer) NS1298-1-1880.

photograph, had been supplied to Laird by Benjamin Sheppard, a prolific and well-known photographer from Sandy Bay in Tasmania.

Norman Laird was a photographer, conservationist and film maker. In charge of the Tasmanian Government's Film Unit from the late 1940s, he was also a dedicated thylacine enthusiast whose collection of photographs, records and research papers were gifted to the Tasmanian Archives by his family in 1990.

The photograph in question turned out to be pasted into one of Laird's research books and came from an article in the 16 August 1956 copy of the *Australasian Post*. Laird's notes state:

> The photograph in this item was taken by, and is the copyright of Ben Shepard [sic], Photographer, of Sandy Bay, Hobart. The negative is a glass ½ plate, with much knife work on it to reduce contrast and background. A print was purchased by me for 3 pounds, and permission was given to use it in a book publication. The animal died the day after it was photographed, and it does not represent the species in good condition, in fact, the subject of the photograph is emaciated. NL (Laird no date).

Benjamin Sheppard is best known as a photographer in Hobart. He worked for the *Tasmanian Mail* and *The Mercury* and as a private portrait photographer from his home in Sandy Bay. Among the photographs of family, scenery, weddings and portraits held in our collection, were three versions of the Sheppard's thylacine image, two were close-ups of the third (NS1298/1/1880). The date on the archive's record for the photograph was given as May 1936.

The question for us was where did this date come from? We needed to be sure because as Gareth explained, if this was the actual date it could very well be the last known photograph of a living thylacine.

The Benjamin Sheppard collection also contains his appointment diaries for the years 1921–73, as well as three general photographic workbooks. My first step was to comb through his appointment diaries to see if anything mentioned the Beaumaris Zoo, especially a thylacine in the relevant years. After drawing a blank in the appointment diaries, research moved onto the workbooks. Among the entries in the workbook for 5 × 4 inch plates, under a column date 'May 1936', was 'Tasmanian Tiger, Beaumaris Zoo' (Fig. 25), which along with references to a 'white oppossum' and a 'Forester Kangaroo', are further bracketed next to 'May'. As a further verification, the workbook gave a reference to where the original negative was stored, 'Box 25'; fortunately, Box 25 is also held by Libraries Tasmania, within which was found the original negative of Sheppard's thylacine image. The page from Sheppard's workbook was then digitised and along with Sheppard's image, and is now listed on the Libraries Tasmania website with the correct date, and provenance.

As a result of this research, the longstanding confusion regarding the image's date, and the identity of its subject, has been resolved. Although other images of the last captive thylacine from 1936 have come to light since 2018, none is later than Sheppard's, which remains the latest confirmed document of a living thylacine.

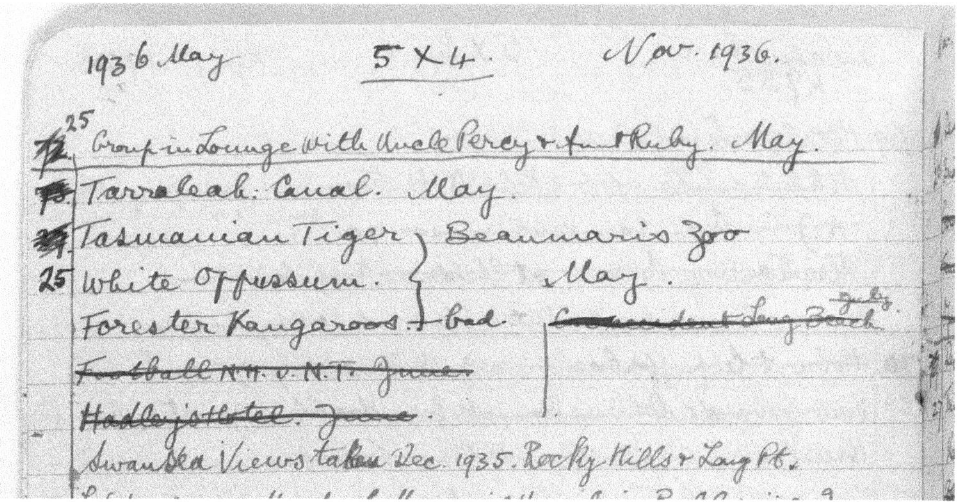

1936 May 5 X 4. Nov. 1936.

25 Group in Lounge with Uncle Percy & Aunt Ruby. May.

Tarraleah: Canal. May.

Tasmanian Tiger } Beaumaris Zoo

25 White Oppossum. } May.

Forester Kangaroos. bad.

Swansea Views taken Dec. 1935. Rocky Hills & Long Pt.

Fig. 25. Extract from Sheppard's 1936 workbook. Tasmanian Archives: NS1298-1-1880.

A lame and lonely creature: the end of the photographic record

Gareth Linnard

Unsurprisingly, the last captive thylacine was the most photographed, with 14 still images published by Sleightholme and Campbell (2021a), a further two images recovered in 2021 and 5 film sequences of between 7 and 54 seconds (Sleightholme and Campbell 2021b). It is through this photographic record that the last captive is most widely known.

Linnard *et al.* (2020) propose an additional film sequence also captures this individual. On 23 December 1930, 2 days after his ordination at Hobart's Saint David's Cathedral (*The Mercury* 1930b), Reverend Harold Doyle visited Beaumaris Zoo with his Pathé Baby cine camera and filmed just over 5 seconds of footage of a juvenile male thylacine (Fig. 26). Though its status as the last captive is presently contested, this can only be the Delphin animal. The Doyle film captures a young animal energetically running back and forth across the field of view, and, controversy aside, is the only recovered film sequence of a subadult thylacine.

The only other known film of the last captive thylacine as a juvenile was taken between 26 March and 17 April 1931 (*The Mercury* 1978; Sleightholme *et al.* 2020), by honeymooning couple Dorothy and Dr Randall Stewart, 3–4 months after the Doyle film was shot. Frustratingly, despite being available in 1978, when a copy was sent by the Stewarts, then resident in London,

Fig. 26. The young Delphin male [still image from 1930 Doyle film]. Courtesy: John Doyle.

to the Tasmanian Broadcasting Corporation, all copies of the film have since vanished. The only remaining fragments being two stills published in *The Mercury* (1978).

With a running time of 45 seconds, Fleay captured what is universally regarded as the most famous moving footage of the last captive. In addition, he took six photographic images. A well-known point of pride for Fleay was that he sustained a bite to the buttock from the thylacine while filming: 'The big fellow in the zoo was not a safe companion inside his enclosure, and while photographs were being taken Mr. Reid had to ward him off continually with a paling' (Fleay 1934).

A further cinematographer who visited the thylacine enclosure was Fox Movietone cameraman Erik Bierre, in 1934 (Lansley 1978; Sleightholme *et al.* 2020). Bierre recalled: 'Two interesting things I filmed at the zoo at Hobart. One was a Tasmanian tiger. He was quite harmless. Very much like a big dog, about the size of an Alsatian, short haired and striped, with a most extraordinary useless thick tail. Like a kangaroo's tail except a kangaroo uses his tail for balance, this thing as far as I could see was just a nuisance. But it was a great thick tail. Where it joined the body it would have been about four or five inches thick, and tapered down, and it was fairly long' (Lansley 1978). Bierre's reference to the girth of the base of the tail is suggestive of the penile mound.

Bierre's recollection, evident in the film (Fig. 27), suggests that the thylacine was quite relaxed in his presence. This appears to be in contrast to the somewhat agitated behaviour seen

Fig. 27. Tasmanian Tiger – Thylacine [still image from the 1934 Bierre film]. Tasmanian Archives: Agent General's Office, photographs, AA193/1/1003 c1930.

in the Fleay film, together with the bite, or nip, he sustained. However, as former wildlife film producer, Martin Banks, notes: 'Fleay's jerky film always gives the impression it was hesitant/nervous, but it may have been the film speed, rather than the animal' (M. Banks, *pers. comm.*). A slower frame rate creates a different, more natural impression of the last captive, as can be seen in the enhancement of Fleay's footage by Composite Films, commissioned by the National Film and Sound Archive to mark the 85th anniversary of his death.

Though a further two undated sequences exist, the latest dateable film of the last captive was captured around March 1935 by Brisbane cinematographer Sidney Cook. Although the Cook film was made only 18 months before the last captive's death, there is no discernible alteration between the subject's appearance and that visible in the Fleay or Bierre's films. Yet, subjectively, a close examination of the penultimate still image, taken nearly a year later by Scottish engineer John McDowell, in January 1936, gives the impression of a less robust animal (Fig. 28). His skeleton appears prominent in some areas and his muscle mass appears less. Between 3 and 4 months later, a photograph taken by Benjamin Sheppard in May 1936, the last known image of a living thylacine, appears to show a further deterioration.

If Doyle's film captured a lively young animal and Fleay's and even Cook's films captured a 'fine male thylacine' (Fleay 1934), Sheppard's photograph depicts the 'lame and lonely creature' recalled by Sharland (1962). Sheppard's image of the last known thylacine, in apparently poor condition, alone in its stark concrete-floored enclosure, suggests both a prolonged decline in the animal's health prior to its death, 4–5 months later, and an ignominious end to our connection with the species.

Fig. 28. McDowell image: the penultimate known image of a living thylacine. Tasmanian Archives: AA193/1/1003.

An ethereal appearance: the last known living thylacine

Gareth Linnard

Following the death of the penultimate captive at London's Regent's Park Zoo, on 9 August 1931, no further thylacines were recorded except for the young male in Hobart's Beaumaris Zoo

Unfortunately, the last captive thylacine is still viewed by many as an animal considered unremarkable during its lifetime, that was allowed to die of neglect and whose body was casually disposed of, none of which is supported by evidence. This is something of a modern indulgence.

Accusations of disregard and deliberate neglect are poorly founded. On 13 May 1935, the Town Clerk, W. A. Brain, wrote to Bruce Lipscombe, in administrative charge of the zoo after Reid's death in December 1935, requesting that the thylacine's hours of display be reduced and that it should be the first animal to be put away in the evening (Paddle 2000). After the last captive's death, the HCCRC attempted to replace it at £30–40 (HCCRC 1936a; *The Mercury* 1937b), hardly the reaction to the loss of a disregarded animal, especially during a depression, or one whose value meant it was likely to have been wilfully neglected.

Had the last captive's death resulted from an act of neglect, the zoo's opponents, who were rather numerous by 1936, would justifiably have made a great deal of political capital from it, yet they did not. Within 6 months of its occurrence, references were made to the last captive's death by the zoo's two most committed critics, political activist Edith Waterworth and Alderman Samuel Crisp, both of whom understandably objected to the sum of public money the zoo consumed. Albeit a little flamboyantly, Waterworth (1937) wrote: 'The frozen despair which its face and its whole body expressed would wring the heart of any person not entirely without imagination. Its frenzy was over, it had refused to eat, but the sight of it will live in my memory to the end of my life'. Crisp wrote that the thylacine had 'fretted itself to death' (*The Mercury* 1937b). Although neither is explicit nor written in objective terms, there is a notable similarity between the two references. Both appear to suggest a prolonged period of decline in the last captive's condition prior to death, supporting the subjective impression created by the 1936 images, though there is no more we can definitively say about the cause of the animal's death.

Although a single act of neglect is unsupported, concerns over Beaumaris Zoo's care of its stock in general are well documented, yet so is praise, though the latter was less common during the 1930s. After meeting the last captive in December 1933, Fleay (1934) publicly described him as a 'fine male' that had 'thriven' at Beaumaris. Privately, he appears to have been less impressed by Reid's husbandry, and would later write: 'How I longed to get this evidently hungry animal across the Bass Straight for specialised housing and feeding' (Fleay-Thompson 2007). Fleay also provided what is probably the most evocative description of the last captive: 'Still wearing the springer brand snare brand around his right hind leg, this long, lean, softly padding animal had an ethereal appearance. He regarded me incuriously, as ceaselessly on the move, he halted now and then to indulge in the widest yawns I've ever seen. The distinctive darker cross-bands decorating mid-back to tail-base stood out against the olive brown coat colour'.

Aside from the comments of film makers such as Fleay and Bierre, very few contemporary references to the last captive survive. The *Advocate* (1935) reported that the thylacine was fed '3 lbs of meat per day' and on 27 January, *The Mercury* (1933) reported the comments of naturalist Alec Chisholm: 'The quarters of the thylacine (Tasmanian tiger) could be improved; he objected to animals being kept on concrete'. A brief comment by Reid was reported in the *Illustrated Tasmanian Mail* (1934) in December: 'The two principal attractions in the zoo said Mr Reid, are the marsupial wolf and the Tasmanian devil. The wolf is one of the oldest types of mammal, and he has a reptilian tail he cannot wag. He is the only one in captivity in the world, and we could sell him anywhere, but I would not like to see him go. The wolf does not live to a great age, and is not a prolific breeder'. Beyond these, only very brief contemporary references have been recovered and they do no more than acknowledge the species' presence at the zoo.

When, on the evening of 7 September 1936, the last captive thylacine died, it is sometimes commented upon that there was no immediate mention of it in the local press, but little should be read into this. Zoos do not tend to advertise the death of their stock, as it is not a point of pride. The first *privately* circulated reference to the death appeared just over a week later in the minutes of the next meeting of the Reserves Committee, held on 16 September 1936: 'The Superintendent of Reserves reported that the Tasmanian tiger died on Monday evening last, 7th instant, and the body had been forwarded to the Museum'. Under this entry, in capitals, is written: 'Noted. Efforts to be made by superintendent to obtain another tiger up to value of £30 each' (HCCRC 1936a). Thankfully, for any thylacines potentially remaining in the wild, by the time this decision was made, the species had received full legal protection and was safe from incarceration in the marsupial carnivore enclosure at the rear of Hobart's declining zoo and the offer to purchase was retracted when the Fauna Board threatened to prosecute the council if they persisted in attempting to procure a replacement.

Despite it being unambiguously recorded that its body was forwarded to the Tasmanian Museum, to the best of my knowledge, no physical remains of the last known thylacine can be traced in the museum's collection today. Perhaps its skin, with its distinctive zig-zag stripe pattern, as can be seen in the Fleay film and Sheppard photograph, was considered too degraded to be worthy of preservation. Yet, while the skin may not have been preserved, the rest of the last captive thylacine was a valuable specimen and it is entirely possible that some part, or all, of the skeleton rests unrecognised in boxes at the storage facility of the Tasmanian Museum. The facility is located just across the River Derwent from the former site of the Beaumaris Zoo, and of the tiny enclosure, roughly the same size as a standard car parking space, which the thylacine patrolled for the last 6 years of his life.

The trouble is to catch the beggars: 8 September 1936

Gareth Linnard

> Are we going to sit down quietly and allow this unique animal to die out within a short period? Something must be done and done quickly, if we wish to avoid the slur which is bound to fall upon us by its extinction. If every interested

> person, and everyone should be interested, voiced his and her opinion we might get a move on in the right direction, and win the approbation of the world for our efforts in saving the Thylacine (Colbron-Pearse 1937).

The most significant action to determine whether the thylacine still persisted in the wild in the late 1930s was undertaken by Tasmania's Fauna Board. The commencement of their endeavours, just 6 months after the death of the last captive, could potentially give the impression that this motivated them to action. In fact, neither the Fauna Board's undertakings nor the questions they posed themselves were prompted by the death of the last captive. Instead, they were, initially at least, instigated by Cecil Godfrey Ryan, mining engineer, entrepreneur and cousin to Albert Sherborne Le Souëf, Director of Sydney's Taronga Park Zoo. Around early February 1937, Ryan wrote to Tasmania's State Premier, Albert Ogilvie, regarding the potential profitability of breeding captive thylacines and Ogilvie forwarded the letter to the Fauna Board where it was read at a meeting on 9 February, after which, as reported in *The Mercury* (1937a), there was a discussion between the Chairman, Colonel John Lord, and member A. W. Burbury (representing the interests of the livestock owners on the board and not the thylacine's greatest advocate) of what was essentially the crux of the board's problem in the proceeding:

> Mr. A. W. Burbury: The scheme he mentions is all right, but the trouble is to catch the beggars…We have no reliable evidence of the present existence anywhere of these animals, have we?
> The Chairman: No.
> Mr. Burbury: Then we do not know that the animal is not extinct.
> The Chairman: No – that is our job to find out if there is any place where a tiger has been seen of recent months, even of recent years.

Two 1937 expeditions were initially planned to determine this, with a third in November 1938. Their intention was not to capture, photograph or even necessarily sight a thylacine, but strictly to attempt to find traces of the species' presence. However, before any expedition set out, the Fauna Board's initial task was to gather and collate reports from recent witnesses to the species in order to identify potential survey locations. It appears that a request was broadcast via the Australian Broadcasting Commission (*The Mercury* 1937c) to gather reports, at least 17 of which were received by the Fauna Board between February and September 1937 (Sleightholme and Campbell 2016). Despite this, the Fauna Board seem to have decided by 24 February 1937 on the informants whose testimonies the expedition locations were selected: Roy Marthick of Smithton and Leslie Williams of Cardigan. Or at least, they had already heard from them by then, as a memo from Colonel Lord to superintendents Oakes of Hobart and Grant of Burnie, confirms (Lord unpublished c).

The first expedition comprised Sargent M. A. Summers of Wynyard, Trooper G. G. Higgs of Ulverstone and Marthick. The expedition set off on 23 April 1937 to search an area west of Mount Bischoff, bounded by the Arthur River to the north and the Pieman

River to the south. Lord subsequently described the results as 'rather negative' (*The Mercury* 1937d). The second expedition, led by Trooper Arthur Fleming accompanied by Williams, departed on 28 November 1937 to search an area south of the Lyell Highway to the west of Frenchman's Cap. Fleming, an experienced bushman, reported to Lord that he had identified thylacine tracks in 11 places (*The Mercury* 1937e). A third expedition set out the following year, on 12 November 1938, to search the area to the east of Frenchman's Cap, north of what is now the Franklin Gordon Wild Rivers National Park. On this occasion, the *Royal Zoological Society of New South Wales* (RZSNSW) was granted permission to send an observer, a position filled by Michael Sharland. Sharland published an account of the expedition in the 1939 RZSNSW's annual report. The party was again led by Fleming and, in addition to Sharland, consisted of Trooper Robert Boyd of Fitzgerald and Constable John Royle of Hobart. In his 1939 report, Sharland noted that the party had located tracks in five, possibly six, locations (Sharland 1939).

Despite his optimism in 1939, 24 years later in his book *Tasmanian Wildlife* (Sharland 1962), Sharland returned to his 1938 expedition, but with a noticeable change in tone. What had in 1939 been multiple accounts from the prospectors of the Jane River, some of which were second- or third-hand, had become: 'Of those Jane River men working here at this time – and some had lived in this country for several years – not one claimed actually to have seen a tiger in its native state', and, 'the fact remained that in this equally wild and inhospitable country about 'the field' no Tiger had ever been seen'. In addition, the tracks observed in the five or six locations had contracted to only one, at Thirkell's Creek. These tracks were cast and the copies are now held at the Tasmanian Museum. It is not possible to assess the surviving casts as representative of all the tracks found, nor if they were representative of those found on Fleming's previous expedition. Also, there is no mention of which member of the party identified them. It is, however, possible for the casts to be examined today by someone experienced and qualified in determining the tracks of Tasmanian animals, the results of which may be surprising (see pp. 139–42).

White mice at Hobart: 1936 revisited

Gareth Linnard

Perhaps, reflecting the hesitancy of the author to let go of Hobart's last thylacine, and continue into 1937, let alone 2022, with no confirmed thylacines in it, it seems fitting to consider what remains of the last captive's world and the fate of the zoo that held him. And to consider why, after all, 1936 was a defining year for the species.

On 16 June 1937, the Reserves Committee made a series of recommendations, among which was that the now half empty marsupial carnivore enclosure be demolished, on the grounds that it was 'unsightly and sunless' and its remaining occupants, Tasmanian devils, be moved to a new enclosure (HCCRC 1937). This decision is probably why, although many of the concrete bases of the zoo's enclosures still survive, as anyone can attest who

Fig. 29. The sign from Beaumaris Zoo's main thylacine enclosure. Courtesy: John Doyle.

has surreptitiously probed the site of the thylacine enclosure, and several have, no trace can be found of Reid's infamous concrete floor on which Fleay's ethereal animal once softly padded, along with around one-quarter to one-third of the live thylacines captured after 1923 (Fig. 29). However, not all trace of the last captive's world has disappeared; the polar bear enclosure visible in the background of many photographs still stands, dilapidated but largely intact, as does a tree that once shaded the enclosure's southern perimeter.

A further recommendation agreed at the meeting of 16 June 1937, was the purchase of two 'Gibbon Apes' from Thailand, thereby breaking the committee's own embargo on the purchase of foreign species, outlined in a report submitted on 19 February 1936 (HCCRC 1936b). The ensuing furore ultimately resulted in the council tabling a final, and successful, motion to close the zoo. After evading the municipal axe for the best part of a decade, Beaumaris Zoo, once prophetically dismissed in its planning stage by Alexander Marshall, MP for Bass, as a collection of 'white mice at Hobart' (*The Mercury* 1922a), finally closed in late 1937.

What became of the thylacine in the wild after 1936 is not a question that can be currently answered in any definitive sense, for the same reason that it is not clear what state the wild population was in at any time. A valid response would have to be evidence based and that evidence is deficient. That said, as the most compelling evidence of thylacine distribution is their captures and the least ambiguous evidence of those is often the record of their sale; perhaps the evidence lends itself to a related, and very relevant, question.

For nine decades all evidence of the thylacine has been anecdotal. When did the thylacine transition from a species whose financial value was capitalised on, to one whose existence is based solely on eyewitness reports?

As far as can be determined, unequivocally, the evidence confirms this occurred not in September 1936, but almost 6 years earlier on 30 October 1930, with the sale of the first Delphin animal to Beaumaris Zoo (HCCRC 1930). This is inescapable, yet not quite as open and shut as it may seem. The last thylacine definitely confirmed in the wild was the second

Delphin animal and although the evidence for the Kaine capture in early 1931 is compelling (Sleightholme *et al.* 2020), there is no confirmatory evidence of its sale.

Further, Haygarth (2017) demonstrates that the capture of thylacines was an unintended, but welcome byproduct of the general fur trade. Although trapping was extensive throughout Tasmania, and during the 1930s was unquestionably occurring in the areas from where thylacines had been captured or killed within the previous decade, opportunities for large-scale trapping, which saw incursions deeper into the bush, were limited. During the 1920s and 1930, of the thylacines either killed or captured by snare or trap, and for which there is a recorded or well-supported and specific date, all but three were definitely procured within the yearly game season. One of the remaining three, the Jordan/Gleeson thylacine is uncertainly dated and may in fact have been taken in the game season of 1928.

In 1929, the decision whether or not to open the game season passed from the State's Attorney General, who as an elected official was always subject to political pressure for it to be opened, to the newly instigated Fauna Board, headed by Colonel John Lord. Free from the ire of the state's voters, the Board immediately closed the game season for 1929, opened the season fully in 1930 and 1931, closed it in 1932, closed it for possum but opened it for wallaby in 1933, opened it for possum but closed it for wallaby in 1934, closed it completely in 1935 and opened it for only 2 months in 1936.

There can be no doubt that thylacine captures correlate strongly with widespread fur trapping; specifically, the trapping of wallaby, as possums were more usually shot. Given this, the likelihood of a thylacine capture was significantly raised only in 1930, 1931 and 1933, the short 1936 season coinciding with the species being granted full protection. That no thylacines were captured in these game seasons is far too inconclusive to suggest that it represents the species' extinction in the wild. This somewhat mitigates the otherwise fairly damning fact that despite there being eager buyers offering £20–25 for a live thylacine, and £10 for a body, not one was presented. However, inevitably it is justifiable to say that by this time the species had become undetectable.

June 1936 saw the introduction of full legal protection and it is in this legislation that the year's most significant implication lies. Not because it offered the thylacine any actual protection, it did not, but because it lessened the likelihood of a capture being reported. Though thylacines were specifically sought, dead or alive beyond this point, on behalf of both the Tasmanian Museum and David Fleay, no-one, without the special prior permission of the Board, could legally profit from their sale. The motive for presenting thylacines was always profit. In the 21st century this is condescendingly seen as human nature, but in the context of life in rural Tasmania during the 1920s and 1930s it was necessity.

Resultantly, any potential captures that were not pre-arranged are highly unlikely to have been presented after this time, as their captors stood to gain nothing and risked prosecution. Though again there is a caveat: although the chances of solo efforts seem slim, Harrison had been appointed by the Fauna Board in 1934 to procure thylacines and there is no evidence this permission had been revoked in 1936. It seems unlikely this would have required Harrison to personally retrieve them, so it would appear that he could still have relied on his connections with the trappers of the north-west; yet despite the sums offered in the mid-1930s, no thylacines were forthcoming.

By June 1945 the Fauna Board offered an amnesty against all accidental kills or captures and offered to pay all reasonable expenses for anyone who provided one to the Tasmanian Museum (*Advocate* 1945), yet still none was provided. Notably, 1945, the year the Second World War ended in the Pacific theatre, appears to have heralded a significant increase in interest in the thylacine. In all other years from 1920 to 1954, the search term 'Tasmanian tiger', the most popular colloquial term for the thylacine, retrieves an average of 15 articles per year from Libraries Australia's Trove portal – in 1945 there are 174. Although the ending of the war is unlikely to entirely account for the disproportionate amount of attention in 1945, it should be factored into any consideration of the reaction to the inability to confirm the species' presence in the wild in the late 1930s. Immediately after the Fauna Board's expeditions of 1937 and 1938, the majority of all of the state's resources, attention and priorities would inevitably have been redirected to the war effort.

Yet the implications of the overall evidence are still sobering. There is no definite contemporary evidence of the existence of the thylacine to the east of the fringes of the settled districts in the western third of the state after June 1920. In the state's south-western third, trapping was both intense and had definitely penetrated the remote districts, particularly in 1933; despite this, after 1926 there is no evidence for thylacines being traded from this area. In the north-west, during the early to mid-1930s, thylacines had been retrieved from the Arthur River, but contemporary reports cease at 1930.

As to the remainder, an area roughly between the Arthur and Gordon Rivers, south of which was well trodden by the trappers of the south, as Sharland describes, this area intersected by the Lyell Highway was not virgin territory and was visited by snarers and prospectors, none of whom is likely to have forgone a stripey windfall should it have appeared in their snares. Yet, again, there is no evidence for thylacines being traded from this area after the 1920s.

Though there is no certainty that it had not, this is not intended to be an argument that the thylacine became extinct after the early 1930s, because the evidence is too ambiguous. Rather, it is an acknowledgement that despite being valuable and sought after, evidence supporting the thylacine in the wild beyond that point is entirely anecdotal, exactly the position in which it is in today.

Plate 1. Head skin of London Zoo's last thylacine (OUM 7942 – Oxford University Museum of Natural History). Courtesy: International Thylacine Specimen Database (6th revision).

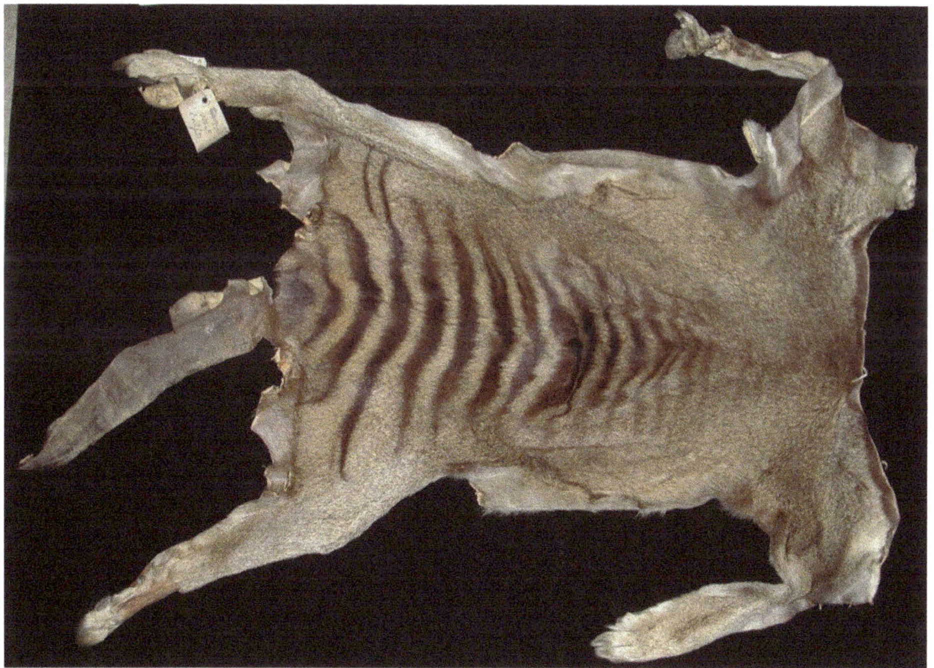

Plate 2. Headless skin of London Zoo's last thylacine (MCZ 36797 – Harvard Museum of Comparative Zoology). Courtesy: International Thylacine Specimen Database (6th revision).

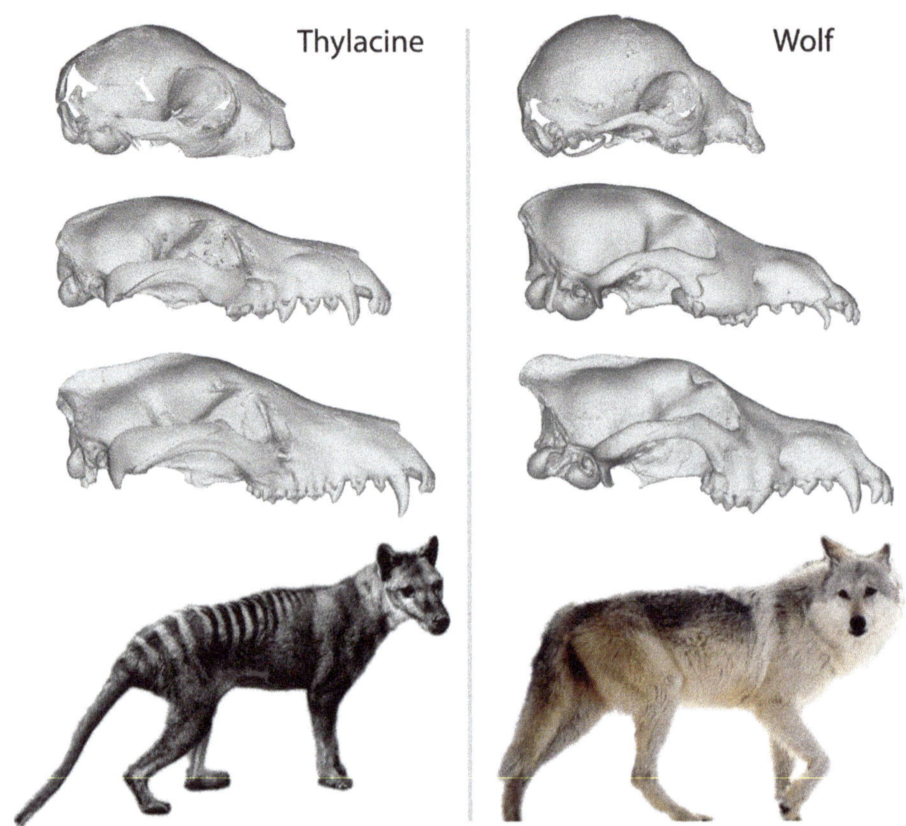

Plate 3. Comparison of thylacine and wolf skull development. Top to bottom: Neonatal, juvenile and adult skull of thylacine versus wolf, showing adult morphology. Similarities in skull shape can be digitally quantified throughout development. Images used with permission under CC BY 4.0 licence.

Plate 4. Endoparasitic larvae of the burrowing flea (*Uropsylla tasmanica*) (above); adult burrowing flea (*Uropsylla tasmanica*) (below).

Plate 5. Adult devil tapeworm (*Anoplotaenia dasyure*) (left); devil tapeworm metacestode (*Anoplotaenia dasyure*) (middle); pigeon roundworm (*Ascaridia columbae*) (right). Tapeworm images courtesy of Prof. Ian Beveridge.

Plate 6. Male devil denning with three females. Photo: Nicole Dyble.

Plate 7. Male devil and sun basking female by Tasmanian artist Nicole Dyble.

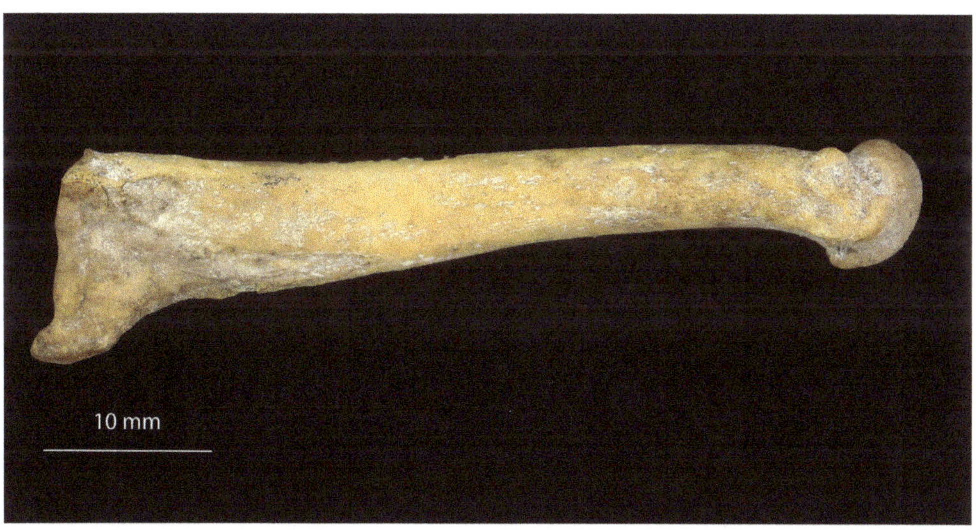

Plate 8. Lateral view of the right metatarsal III of *Thylacinus cynocephalus* excavated from 104.5 to 111.5 cm deep in Caladenia Cave associated with a radiocarbon date of 3254–2925 cal. BP. (WAM 13.11.364)

10 mm

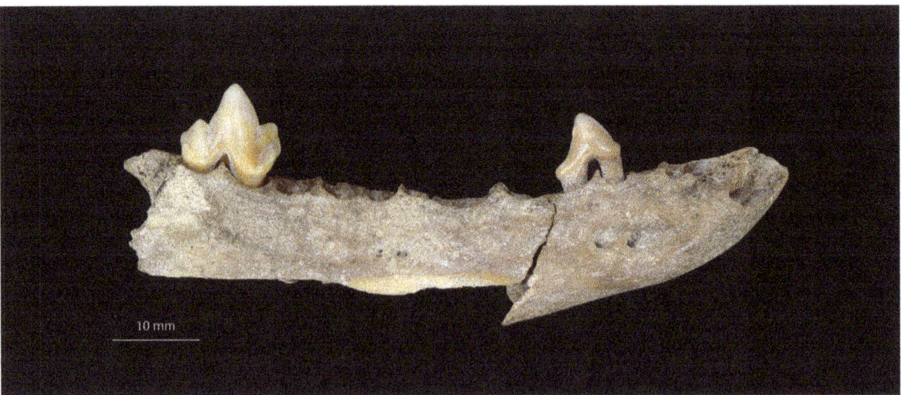

Plate 9. Labial view of the right dentary of *Thylacinus cynocephalus* recovered from 140 to 144 cm depth in Caladenia Cave with an associated radiocarbon calibrated date of 3895–3641 cal. BP. (WAM 13.11.648)

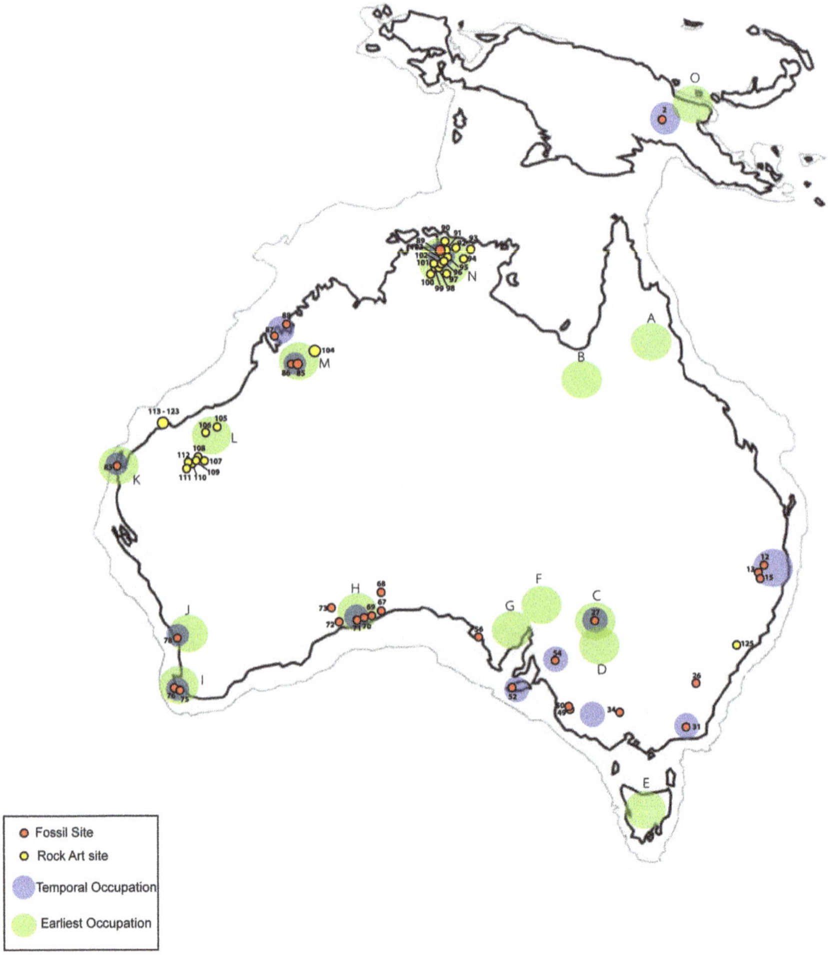

O
N
A
B
M
L
K
F
C
G
H
J
D
I
E
Fossil Site
Rock Art site
Temporal Occupation
Earliest Occupation

Plate 10. Overlap between dated thylacine material and thylacine-related oral histories.

Plate 11. Meadstone thylacine 'lair' in 2001. Author's collection.

Plate 12. 'Dilger Flats', close to where Dilger's thylacine was snared, photographed 100 years on in 2012. The area has been subject to forestry operations for many years. Photo: Tammy Gordon.

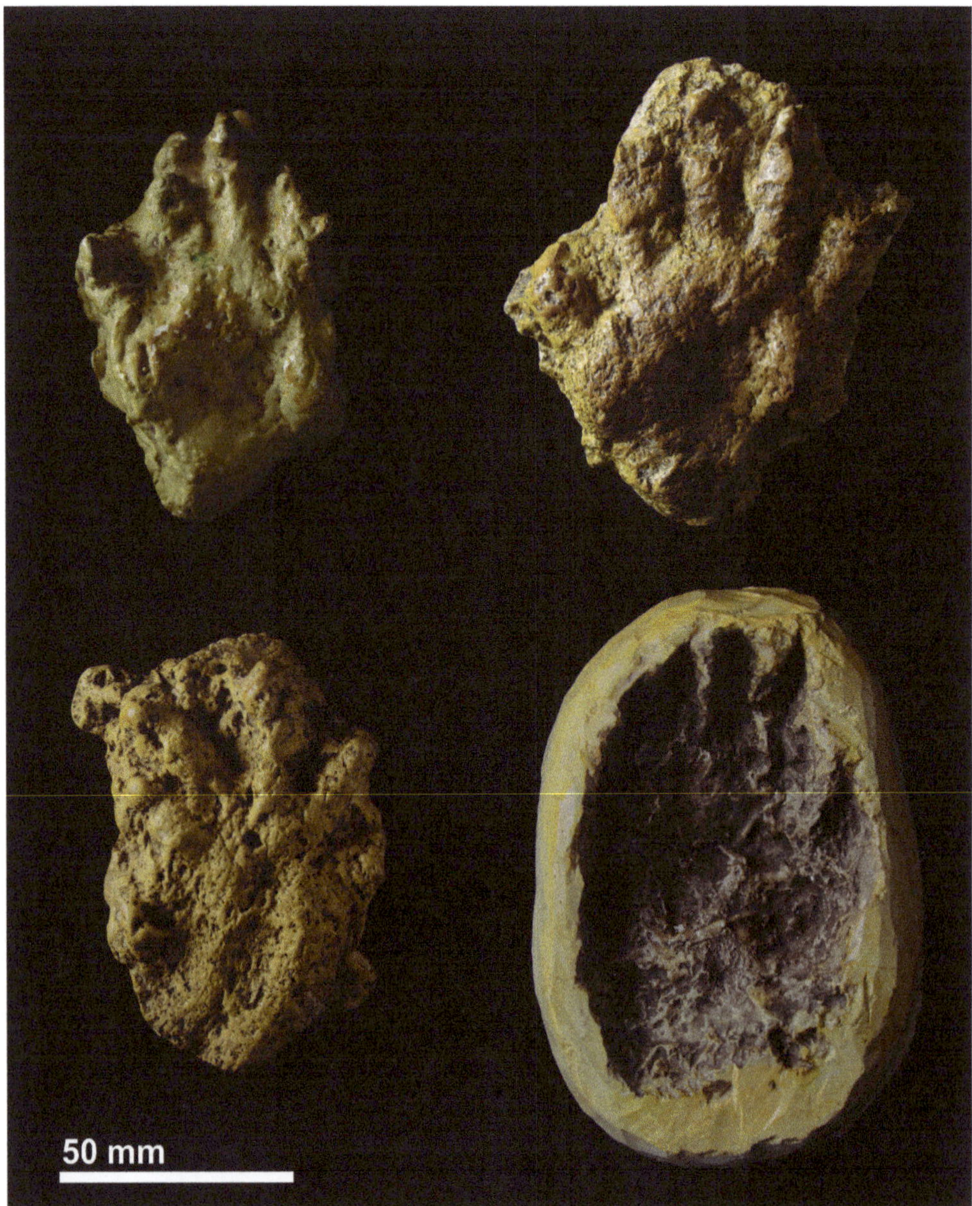

Plate 13. Footprint casts from the 1938 Fleming expedition, taken from the area around Thirkell's Creek, western Tasmania, in November 1938. Photo: David Hocking (Tasmanian Museum and Art Gallery).

Plate 14. The complete Mundrabilla thylacine mummy in 1985. Photographer: Alex Baynes, WA Museum.

Plate 15. Close-up of the head and upper body of the Mundrabilla thylacine mummy in 1985. Photographer: Alex Baynes, WA Museum.

Plate 16. The current display when illuminated by LED lights. Photographer: Clay Bryce, WA Museum Boola Bardip.

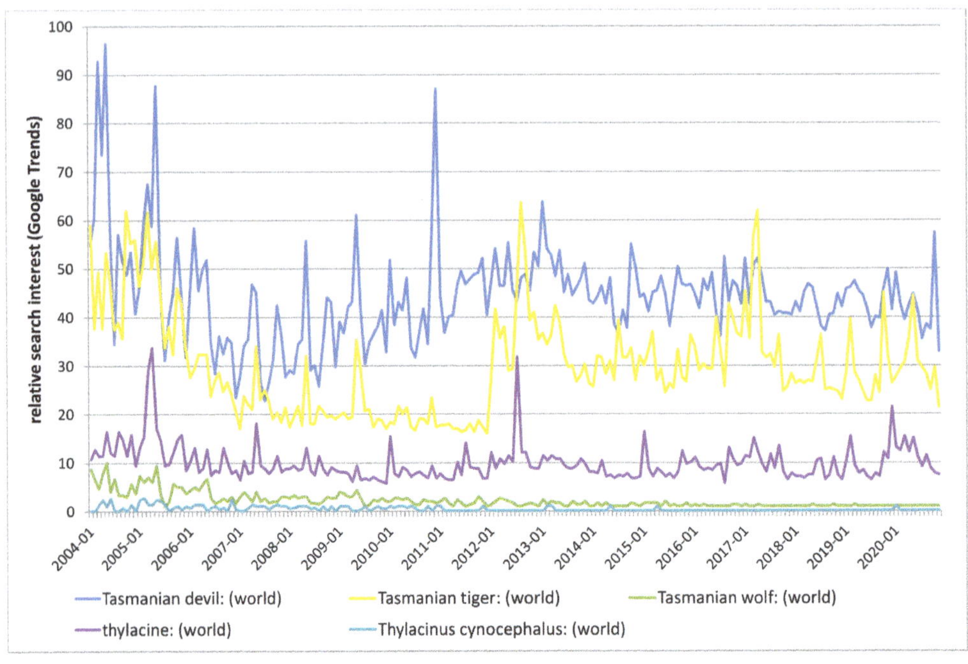

Plate 17. Representation of the popularity of search terms over time (2004–20) worldwide according to Google Trends [accessed: November 2020]. The queries were carried out on at least three different days and averaged.

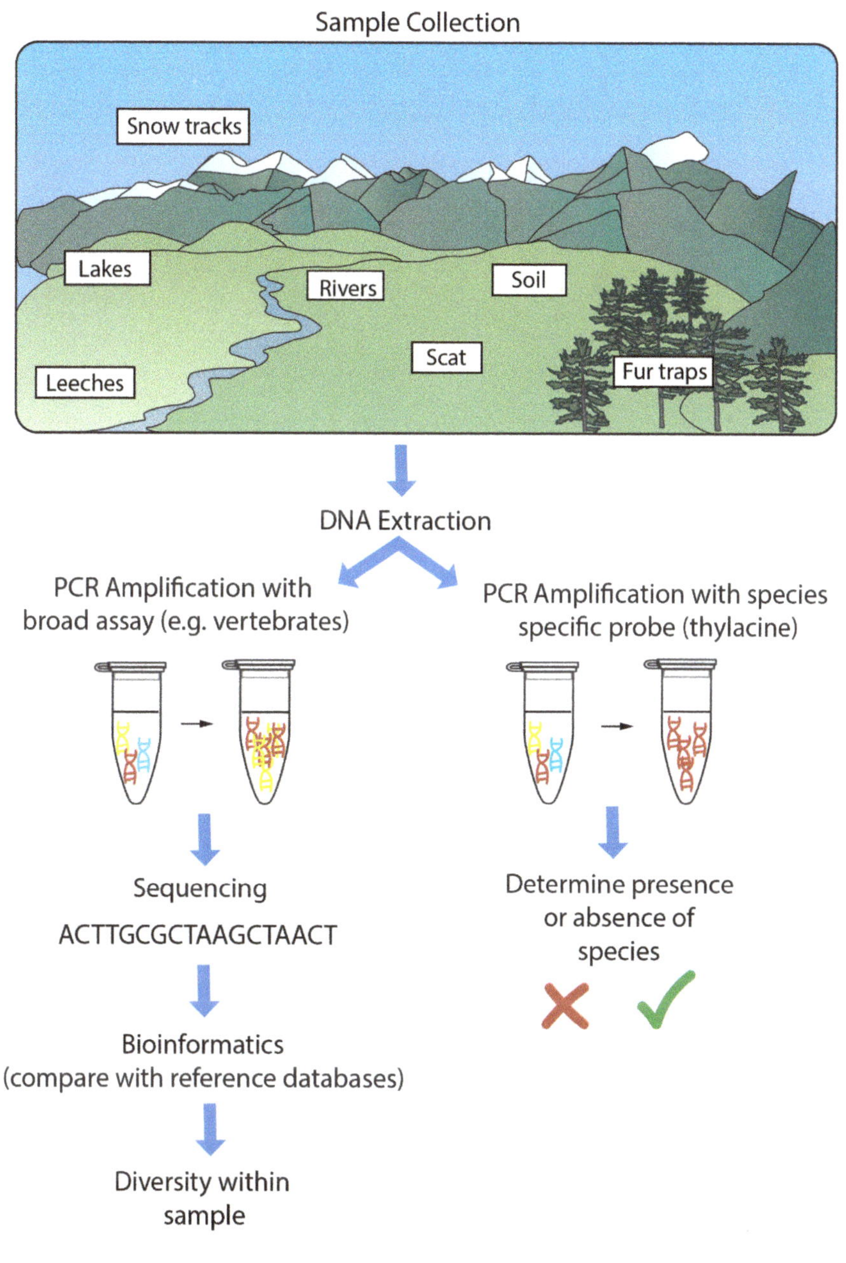

Pros	More diversity data	Quicker, often cheaper
Cons	Cost Rare species may be missed	Development of probe can be costly Difficult to field test probes for cryptic species

Plate 18. Workflow depicting how environmental DNA may be used to detect the thylacine. PCR, polymerase chain reaction. Some elements provided courtesy of the Integration and Application Network, University of Maryland Center for Environmental Science (ian.umces.edu/symbols/).

Plate 19. Extinct and endangered, by Tasmanian artist Nicole Dyble.

PART 7: INTO THE SHADOWS (1937–PRESENT)

Review of footprints from the 1938 Jane River expedition

Nick Mooney

I have examined four synthetic copies of plaster casts held by the Tasmanian Museum and Art Gallery: three casts of footprints ('male' copies) and one (indented) reproduction of a footprint (a 'female' copy). The prints will be discussed under the subheadings of top left, top right, bottom left and bottom right (Plate 13). They were otherwise not usefully or consistently labelled, though they are easily distinguishable.

Basics of thylacine footprints

My thylacine 'gold standard' for comparative purposes is the photographs of cast reproductions from a fresh dead captive immature thylacine obtained in 1920 and held currently in a collection in Victoria. The only visible abnormal feature is the claws, which are very worn on a flat plane, evidence of being kept on concrete, as was usual for zoos in those days.

Superficially, the differences between the fore- and hindfeet are similar to other quadruped pounce–pursuit predators, with the usual standing forefoot ground contact having a lower aspect ratio (more rounded) than the hindfoot. The forefeet are used by the animal to steer, pin, manipulate and brake. The hindfeet are slightly relatively longer and likely more rigid becausee they are mainly used to push the animal forward. Such differences between hind- and forefeet are widespread between taxa and are very apparent in dogs and cats, quolls and Tasmanian devils.

The toes of the thylacine's forefeet are markedly different in size, those of the hindfeet less so, reflecting the more diverse roles of the forefoot. In deep substrate the forefoot of a thylacine would show five toe imprints (as with devils) but not so with the hind. For both fore- and hindfeet of the thylacine the claw tips transcribe an arc with no toe or claw standing out as remarkable.

Each foot's plantar pad has distinct segmenting, the hind with clear longitudinal trilobate sectioning, more so than in Tasmanian devils' forefeet but less so than in quolls, especially the spotted-tailed quoll (which has three longitudinal climbing pads on the hinds). Clear whole or near-whole footprints of a thylacine could thus be differentiated from those of devils by the structure whatever the size (it must be remembered that juvenile thylacines would be much smaller than large devils for some months).

Assessing the museum footprints

None of the prints can be attributed to dogs, given the large plantar pad/single digital pad ratios (~7/1 vs about 3/1 for dogs and cats).

There is no information on whether the footprints were found together or separately nor whether they were mixed in with those of other species. Therefore, I will treat each print individually and on its own merits.

All of the prints are in the possible size range of free-living wild thylacines (i.e. from 3 to 35 kg). They are all taken from deep substrate, probably mud. Several are incomplete as cast, possibly because rocks or other debris took some weight of part of the foot. Such permeations are common with animal footprints.

Top left print (Plate 13)

The outstanding features of this print are the apparent weight on the front part of the plantar pad, signs of flexible toes and apparently short claws. The print is very similar in structure to that of a large devil, although its component areas are larger by 40–50%. The lack of sizeable claws largely eliminates wombat, although if sudden weight on a wombat forefoot is taken solely on the pads and toes, angling the latter upward, there will be no or very little impression of claws. However, in this print there are short claws, so it is from a devil or a thylacine. In my opinion the lack of an angled edge on the front of the plantar pads makes it unlikely to be a devil, no matter the size.

I think on balance of the characters, this is the right forefoot of a medium-sized thylacine; the inside toes and rear parts of the plantar pad are not showing clearly, so the animal was probably stepping down and taking weight on the front of the foot.

Top right print (Plate 13)

This print is superficially similar to the right forefoot of a wombat. However, a wombat's fore toes are relatively short and the claws relatively long. Though it is possible that error led to the claws being missed during casting, this print shows no evidence of claws. Given the rim of the cast, one would expect at least some claw indentation if the print was that of a wombat, especially on what is the top left toe. Although a devil's forefoot can splay similarly, when doing so the thumb typically splays most significantly, which does not happen with a wombat, and I suggest that thylacine and devil forefeet are similar albeit the thylacine's are more rigid, as is somewhat suggested here.

The print is 70–80% larger in area than that of a 9–10 kg devil, which means its 'owner' would be nearly 17–18 kg (the ratio of pad area to weight of quadrupedal carnivores is remarkably consistent across species), far too large for a devil but very much in the thylacine size range.

I think on balance of the characters displayed in this print, it is of the left forefoot of a medium-sized thylacine, albeit with the inside toes and rear of the plantar pad not showing clearly. Spreading the toes is a normal reaction to stepping in deep mud.

Bottom left print (Plate 13)

This is the least informative of the casts and was made by a smaller foot than was responsible for the other three print casts. It is again a forefoot, with toes spread wide. There are similarities with the forefeet macropods, but they have large claws that should have been evident to the caster and therefore recorded. However, a miserly amount of plaster has been used (a common economy on field trips where a surfeit of prints is found and there is a limited supply of plaster) and it is possible the caster rationed the plaster to record the main features of the foot. Wallabies invariable put both forefeet down at the same time (as part of the 'punting' gait) and no other casts represent the other forefoot.

I have the impression the foot has slipped somewhat during braking. When doing so, all animals spread their toes. The animal making this print is the size of a large devil (9–10 kg), so I cannot decide if this is the forefoot of a large Tasmanian devil or a small thylacine. It is also just possible that it is the forefoot of a very large wallaby.

Bottom right print (Plate 13)

This cast is an imprint and the most useful detail is the four toes in a shallow crescent with a fifth smaller toe stepped down. I do not think this detail represents an overlay (the superimposition of one foot print over another), rather that the five toes are from the same foot. Therefore, it was not made by a thylacine hindfoot and is far too large for a devil. Claws appear to be short, making it unlikely to be a wombat, even though the position of the toes is similar to that of the forefoot of a wombat, as is the wide plantar pad bearing weight.

On balance of the characters, I think this is most likely the right forefoot of a medium to large thylacine, but it just might be of wombat.

Discussion and summary

The expedition members making this collection, Michael Sharland, Sargent Arthur Fleming, Trooper Robert Boyd and Constable John Royle, should have been very familiar with wombat footprints and one can assume the characteristic wombat hindprint would accompany the foreprint of most wombats, thus alerting those who were examining the tracks. Wombats could be found anywhere there were thylacines.

At least Fleming should have been very familiar with thylacine (and wombat) footprints and one would assume Sharland (a diligent naturalist) would likely have done research before the expedition if he did not have first-hand experience with those species' footprints.

The three casts most likely to be from a thylacine all seem to be of the right forefoot. I do not know if the expedition members chose the forefeet prints because they were considered to be the most diagnostic of the thylacine or because those were all that were found. In addition, as mentioned before, plaster can be in limited supply on field trips and as such the caster has to prioritise which prints are cast.

I think the four casts represent at least two different animals. Three are likely thylacine and the fourth is possibly thylacine. There is a small possibility that all are wombat because that species' footprint characteristics have many aspects in common with those of the thylacine. Importantly, we also do not know what bias may have been applied in the choosing of footprints to cast. Perhaps the expedition members ignored any footprints that did not fit their view of what a thylacine footprint looks like. We will never know.

The collection year of 1938 is not a revelation, being only 5 years (well within the life span of a thylacine) since the last known wild individual was caught/killed and 2 years since the death of the last captive specimen. In my opinion the likelihood of catching or killing the last thylacine is fanciful, so there could well have been some left in the wild. The expedition was after all, very close in time and space to all the final events.

We also do not know beyond the one imprint (labelled 'Jane River area "tiger" 1938') if the expedition members were sure the prints were of thylacines. Of small concern to me is that the expedition members reportedly commented to the effect that they found thylacine footprints to be common, raising the possibility that they were making common misidentifications (me partly sharing their mistake). Of course, it does not follow that the species was abundant; the expedition may have stumbled across a rare locus of activity. A free-ranging quadruped leaves many footprints every night, which will be visible if ground conditions are suitable. Again, we will never know.

The ghost of Huon Valley: did the Tasmanian tiger briefly haunt the south-west?

Branden Holmes

Since the last known thylacine died in captivity, there have been sporadic, yet persistent reports of its survival in the wild. Most of these are isolated events, but one cluster in the Huon Valley (south-west Tasmania) in early 1949 peaked when 20 men with electric lights and weapons headed into the bush to kill a large striped beast that had been terrorising their small community. Descriptions of the animal varied, but there was no doubt in many residents' minds about the identity of the culprit: a Tasmanian tiger, even though there had only been a few scattered reports of the species from the south-west until the early 1930s (Sleightholme and Campbell 2016) and no reports since.

Messrs A. A. Woolley and B. Thorpe allegedly saw a thylacine chasing a wallaby at Huonville in 1947 (*The Mercury* 1947), while nearby, Mr H. E. Woolley claimed that he had been tracking one for two winters around Glen Huon in 1948 (*The Mercury* 1948). Then, at Ranelagh in mid-February 1949, Mr J. H. Lovell and his dog had a moonlit encounter with an unknown animal striped along its hindquarters. He almost stepped on it before it growled at him and jumped over a fence, his dog's refusal to tackle it leaving no doubt in his mind that it was a thylacine (*The Mercury* 1949a). The phantom allegedly then drove 72 of Mr L. A. Hall's sheep into the Huon River on the 15th of the month (*The Mercury* 1949e).

By March, the reports moved 15 km north to Mountain River at the head of the valley, by which time the phantom had apparently added a 'low-throated neighing' to its vocal repertoire (*The Mercury* 1949b). Mr J. Oates saw the now white-chested animal from 'five feet away' before its 'slow gallop' across a hay paddock, as he and his friend Mr Townsend took up residence at Mr Lovell's, though hairs found on a fence were identified as those of a wild dog by some residents. Luckily, Mr Lovell laid baits for the 'strange animal' from which the neighbourhood's dogs fled, as tellingly the 'baked rabbit, usually preferred by dogs, was left' (*The Mercury* 1949b).

Mrs E. Bennett heard it one night at home and although her husband went out into the night with gun and light, his search was to no avail. A party of up to 20 men, including Mr Lovell, tried unsuccessfully to track it for several nights until 2 am as hysteria gripped the small community. In early April orchardist Mr F. M. Knopp discovered an area of flattened grass where the animal, now described as bigger than a dog, had lain when not terrorising the valley (*The Mercury* 1949c). But if the bedding implied its intention to remain in the area, the very last report was a mile away by Messrs R. Vince and F. Wood while mustering cattle at 7.30 am (*The Mercury* 1949d). The animal had now grown even larger and even stranger, with a bull-necked head and no mention of stripes. And just like that, the phantom faded away as quickly as it arrived.

In its place came Trooper Arthur Leonard Fleming, who had led two official Fauna Board searches for the thylacine, in November 1937 and November 1938, both of which found tracks (*The Mercury* 1938). He also took part in the expedition lead by David Fleay from November 1945 to March 1946 that also found evidence of its survival (*Advocate* 1946). After interviewing witnesses in the valley he concluded that the descriptions, though somewhat variable, pointed to the thylacine and requested to be notified should any further tracks be found that could be cast for a definite identification (*The Mercury* 1949e). None was ever forthcoming.

The likely culprit of the hysteria was a large, brindle dog, given the descriptions of the animal, its jumping, growling and the mass of drowned sheep. Not only are these wholly uncharacteristic of the thylacine, but new research pushes back the last confirmed records of the species in the wild to 1930 or 1931 from the north-west of the island hundreds of kilometres away from these alleged reports. This mass misidentification emphasises just how misunderstood the species was during the mid-20th century and arguably still is today.

Early-adopters and innovators of camera traps in Australia: in search of the thylacine[4]

Paul Meek, Guy-Anthony Ballard, Karl Vernes and Peter John Sabine Fleming

The expeditions to search for the thylacine (*Thylacinus cynocephalus*) in Tasmania conducted between 1950 and 1964 can be considered, with a single exception (see Woodford 2015), the first attempts at camera trapping in the Southern Hemisphere. In 1950, the Hobart *Mercury* newspaper reported a story of an 'anonymous Victorian naturalist' placing cameras with a 'trip-string' along game trails in the hope '...that if a "tiger", or any other animal, in fact, passed that way, it would take its own picture'. Unfortunately, the naturalist sought to remain anonymous; apparently worried his attempts '...might make him appear silly' (*The Mercury* 1950).

Between 1959 and 1964 Dr Eric Rowland Guiler, an Irish-Australian zoologist led several 'Woolnorth Expeditions' to north-western Tasmania in search of the thylacine (Guiler 1985).

4 Reproduced from Meek *et al.* (2015) with permission from CSIRO Publishing.

In 1960, on the second expedition, Guiler and his team trialled a prototype camera-trap comprising a Bolex movie camera attached to a snare cable that was set across a hole in a fence used by animals. The camera produced single-frame black-and-white photographs with a white flash. This expedition was unsuccessful in detecting the thylacine, but did get photographs of Tasmanian wildlife and thereby demonstrated the value of remotely triggered cameras. This subsequently led to further searches for the thylacine. On the third Woolnorth Expedition in 1961, five camera traps were used (Guiler 1985). Each comprised a G.45 aircraft 8-mm movie camera with lighting and a treadle-plate trigger mechanism. Consistent with previous placement, a camera-trap was set below holes in fence lines. On this occasion the team successfully captured more photographs of Tasmanian wildlife, with the notable exception of the thylacine. Another unsuccessful attempt was made in 1966 using a modified version of the G.45 camera (Guiler 1985).

In 1968 and 1972 more intensive but unsuccessful camera trapping surveys for the thylacine were conducted by Jeremy Griffiths, James Malley and Bob Brown, using 25 primitive camera traps deployed over a longer period. Despite the newcomers, Eric Guiler continued his searches in 1978 using 15 custom-built camera traps, built at a cost of A$25 000 (Guiler 1985). This prototype should probably be recognised as the first Australian step towards modern, commercial camera traps. Guiler's devices used pulsed infrared beams and circuit boards, which allowed programming of a delay cycle and included a display that showed the number of events recorded.

Another attempt to rediscover the thylacine was led by Tasmanian National Park ranger Steven Smith, who combined a Pentax 35-mm MX motor-drive camera using 40- or 50-mml enses, a Metz 45 CT-1 flash and a bulk film magazine capable of recording 250 pictures (Smith 1981). To trigger the trap, Smith used a Sick Optik Elektorinik infrared source, 'email' relay and 5 m of cable. Reflectors were set a few metres from the camera and an infrared beam projected across an animal path. When animals passed through the beam the camera was triggered. Despite using lures and attempting to mask human scent with meat-meal and singed bracken fern, Smith's 420 camera images, taken over 7 days, failed to record a thylacine. The dogged tenacity and ingenuity of these early camera trappers helped forge Australian camera trapping history.

The Mundrabilla mummy: conservation and exhibition

Mikael Siversson

In October 1966, a remarkably well-preserved, mummified thylacine was discovered in a cave on the Western Australian side of the Nullarbor Plain. David Lowry, a geologist from the Geological Survey of Western Australia (GSWA), together with his wife Jacoba in a volunteer capacity, made the incredible discovery. The thylacine was subsequently extracted from the cave by a joint team of geologists and palaeontologists from the GSWA and the WA Museum (Lowry and Lowry 1967). This was not an easy task as the specimen was very fragile and virtually impossible to move without bits of skin coming off. The Mundrabilla mummy was not the only thylacine specimen in the cave, which also entombed skeletons and/or mummies of Tasmanian devils, dingoes and other marsupials and placentals, as well as reptiles and birds. The cave was aptly named the Thylacine Hole Cave.

It was soon realised that the GSWA lacked proper facilities to store such delicate and scientifically important fossils. In November 1966, the mummified thylacine and the other vertebrate skeletons from the Thylacine Hole Cave were transported to the WA Museum in Perth on permanent loan (Plates 14, 15) until 2020 when ownership was formally transferred to the WA Museum.

The mummified thylacine was placed in a specially crafted, plywood box, sealed with plastic wrap. To avoid contamination from the wood, the thylacine did not lie directly on the wooden base but rested on inert aluminium foil. The humidity level was controlled by silicone gel sachets. After being kept in environmentally controlled storage for 33 years, the Mundrabilla thylacine was first put on display in 1999 when the Diamonds to Dinosaur Gallery of the WA Museum opened its doors to the public. Placed in a hermetically sealed Perspex case, visitors could view this magnificent fossil specimen but also reflect on the sad demise of the species.

The Thylacine Hole Cave has a high content of halite salt, which probably built up over thousands of years from airborne sea spray moving into the cave when high-pressure weather systems travelled north across the Nullarbor Plain. This phenomenon may explain the extraordinary state of preservation of the thylacine, which lay fully exposed on the cave floor for ~4600 years. Salt is an excellent preservative and inhibits scavenging and bacterial decay. Rather than exposing the fragile mummy to additional preservatives, the curators at the WA Museum argued that nature had already done an excellent job in preserving the animal. Therefore, no further stabilisation was needed other than keeping the specimen in a sealed, humidity-controlled case and managing light exposure.

It is well known that visible light, and even more so invisible UV light, will damage fossil specimens on display. The degree of damage is a function of light intensity multiplied by exposure time. If the Mundrabilla thylacine had been well illuminated for the duration of visitor's hours at the museum during the 17 years it was on display in the Diamonds to Dinosaur gallery, its fur would probably have faded to some degree. Also, light damage is not restricted to fading but may also cause brittleness and other issues. To mitigate the effect of light, the deep Perspex case was lowered into a wooden plinth so that light could only penetrate the case from above. It was temporarily illuminated by visitors pushing a button but most of the time the light exposure remained low.

The WA Museum closed its doors to the public in June 2016 in preparation for the development of a new facility, which took just over 4 years and was a challenging project of preserving and integrating the heritage buildings with the newly built buildings. Before refurbishment of the old Diamonds to Dinosaur gallery could commence, all fossils and mineral specimens had to be moved to the museum's Collections and Research facility in Welshpool for storage. Transporting the Mundrabilla thylacine was a delicate exercise but proceeded without accidents. Once in the storage facility the case was covered with a black cloth to block out light.

Four years later, re-installation of exhibit items at the new WA Museum Boola Bardip again required careful transportation of the Mundrabilla thylacine and its heavy Perspex case. In the Portals to the Past exhibit of the Wild Life gallery, the Perspex case was once again lowered into a plinth with only the top surface directly exposed to light sources (Plate 16). The ceiling light intensity was adjusted and set at ~50 Lux at the level of the top surface of the case. By comparison, bright sunlight exceeds 100 000 Lux.

The stable, controlled environment of the new palaeontological exhibit at Boola Bardip will ensure that future generations can continue to view this magnificent fossil specimen and reflect on the often irreversible damage that we humans repeatedly inflict upon nature.

Finding a marsupial ghost

Chris Tangey

It has been 47 years ago since I started my search for the thylacine. I was young (18 years old) and had little knowledge of recording and cataloguing interviews with those who had sighted the animal and those who had hunted it, let alone keeping photos. As a slight defence it is unlikely that much of that would have survived my 2 years constantly trekking in and out of the Tasmanian bush with just a backpack.

I can't relate it all here but maybe that's a good thing. Most of the life of a 'tiger-hunter' is mundane and even dispiriting, to be honest, with lead after lead having mostly zero results. Dr Bob Brown, who later founded the national Australian Greens party, had conducted a search before me with his friend James Malley as well as Jeremy Griffiths back in 1972. We met briefly one morning but it was more of a 'good luck' than an actual exchange of information.

I was very fortunate to have been able to talk with those who stated that they had hunted the thylacine for a living right up until the mid-1930s, some legally, some not, but all with regret in hindsight. Tiger hotspots all over the state kept being mentioned: Weldborough Pass, Lottah, Goulds Country, Upper Blessington, Waratah, Eddystone Point, the Blue Tier, Liffey, Mathinna, Mawbanna and lots more. All places that I would eventually search.

In the absence of scientific study of the species in the wild, the accounts of bushmen have shaped how many view the thylacine's behaviour. They recounted their belief to me that the thylacine had a distinctive scent, and that its eyes reflected a gold colour. That they are also extremely curious, and that when disturbed they were capable of moving quickly back onto their hindlegs like a kangaroo and taking three or four large hops to create a bit of distance. The old hunters also told me that otherwise the thylacine was not capable of moving quickly, never going at more than trotting pace. They recounted that as the thylacine could not outrun their quarry, they would follow its scent for 20 -30 miles until the animal being tracked collapsed with exhaustion. They would then snap its neck and eat only the offal. The problem in finding a thylacine kill was that Tasmanian devils would quickly eat the evidence. The hunters also told me that a 'natural culling disease' would go through both the thylacine and devil populations about every 20 years, but not at the same time as far as they could tell.

Through an arrangement with the Launceston newspaper, the *Examiner*, I was connected to hundreds of people who claimed to have seen a thylacine. Many of them had told no-one about their sighting for fear of ridicule, a powerful deterrent to this day. Perhaps half of these sightings I identified as mis-sightings, the wrong animal, poor visibility or lack of credibility. But I was left with about 150 sightings I couldn't doubt, including one seen eating seafood.

At Eddystone Point (north-east Tasmania) an old fisherman told me that he gutted a fish one evening, throwing the guts out from his shack, which he saw being eaten by a thylacine 30 minutes later. He managed to get quite a good photograph (or so I was told by his son). While he was happy to chat, he firmly refused to show me the image. Not only did this fisherman have a fear of ridicule but also a fear that any identification of the animal would risk 'crazies' coming up and shooting it dead for fun, a very real concern in those days. He also thought that he and his fisherman mates might be thrown out if authorities declared the area a national park. Indeed, the authorities did declare it a national park anyway and I strongly believe that Mt William National Park was gazetted specifically to preserve the thylacine.

The *Examiner* newspaper later facilitated a much better way of chasing sightings. People would ring an intermediary and then I would talk to the witness, often within 24 hours, sometimes the same day. If their report sounded promising I was often on-site within 2–3 hours. Through this improved system, on two occasions I came very close to getting evidence of a living thylacine.

On the southern edge of Arthurs Lake (Central Highlands), where a creek joins its southern edge, I found myself out in the sleet on a frozen-solid mudflat in the middle of winter. The day before a fisherman had called: 'You the tiger man?' 'Yep', 'Well you better get up here, I saw a tiger and four cubs drink at the lake edge this morning.' This man was a bushman and had no doubt about what he saw in front of him that day for more than 30 minutes. Hoever, he only told me on condition that I kept his identity anonymous.

Sure enough the prints were still clearly there, frozen in time, for now anyway. There was no chance of taking a plaster cast in the icy mud, so I took photos instead with a matchbox for scale beside the mother's prints. When I showed the photo to Dr Bob Green, then Curator of Vertebrates at Launceston's Queen Victoria Museum, he commented strictly off-record: 'believe it or not there are no proper casts of a thylacine's footprints so we can only go on the rather shrivelled up and distorted paws of museum exhibits. But if I was to imagine what one would look like it, would look exactly like *that*' as he placed his finger on the photo.

Probably the closest I got was when a young couple called me from Reedy Marsh, north of Deloraine (north-central Tasmania). They told me that the night before they had seen a thylacine in front of their vehicle for 15 minutes on a dirt road in constant rain. Not a fleeting glimpse, but right in front of them, 'leading' them down the road with a weird swagger, occasionally stopping to look back over its shoulder, but never in a hurry to get out of the way. This was to be a common story I heard in many later sightings on roads. Although thylacines can smell you for immense distances and totally avoid you if they choose, once you're near them in a car they take on the arrogance that only comes with being top of the food chain: I'll move when I feel like it.

I set up camp the next night only to have the tent collapse on me after some screaming Tasmanian devils, in an apparent fight, took out one of the guy ropes as they ran past in the ferns. The next morning the sky had cleared but it was difficult to see any tracks after the constant rain of the previous days. I made my way along a track that led to a dead-end at a hollowed-out tree once lived in by a hermit known as 'Jimmy Possum'.

Making my way back along the track I was surprised to find fresh tracks right up to the edge on the other side of the very shallow creek I'd crossed just 10 min before and which I had checked for tracks before crossing and there were none. Even more intriguing was that the prints reminded me of the Arthurs Lake ones, but with a twist. The rear of the animal had one elongated track that rotated twice more, leaving a mark like a section of clockface with trhee hands. I duly photographed it.

I didn't give the strange track a lot of thought until I came across one of my tiger hunter contacts again about 3 months later. I mentioned the circumstances and showed him the photo of the weird track. With excitement and without hesitation he said, 'it's a tiger, no doubt about it'. And even more incredibly, 'it followed you'. He then recounted several occasions on which he and others had been out for hours looking for thylacines only to find one or even two sets of tracks just 100 m behind them after they had turned back for the day. 'They're very inquisitive of humans and with their amazing scent they can easily observe us without ever being spotted. On the odd occasion we surprised them this is the track we would see, as they have a very rigid body they can't turn quickly like other animals. So they have to quickly pivot on their heel, what you see here is its hind leg turning in three moves, overlapping as it goes. No other animal on Earth makes that mark.'

Are they still around today? I can't say, but I can say, never, ever underestimate the thylacine.

My 1984 search for the thylacine in the state's north-east

Winston Nickols

I have a scrapbook bulging with reports of unconfirmed thylacine sightings, including some from National Parks and Wildlife Service (NPWS) officers in recent years. I had long pondered whether the Tasmanian tiger still existed and on one particular evening at the Mole Creek Hotel I heard passionate stories and reports of sightings. There must be some truth in these I thought, so in October 1984 I planned to spend 1 week in the north-east of Tasmania. Visiting, according to NPWS advice, 'hot' Tasmanian tiger locations I would go to four likely sites and try out my plan.

My overall plans were to park and hide my vehicle in a likely area, secure bait bags made of wire mesh containing meat and bones in likely spots on animal pads, install wireless microphones near these bait stations and attempt to sleep during the day and be alert all night. My vehicle was fitted out with a bed, food, two-way amateur radio communication equipment and dedicated sound-recording gear.

I had built two self-contained and weatherproof FM wireless microphones and these would be placed near each bait station around 200 m apart. At my vehicle these signals would be received by dual receivers coupled to a two-channel sound-activated recorder. A third microphone was placed close to the vehicle. I had also constructed a combination camera/light unit. A camera with tele-lens and cable release was mounted on a powerful, self-powered ex-army spotlight – a completely portable assembly that could be easily carried. Exposure and focus settings were preset. In my vehicle I would remain quiet and at the first sound of action near the bait I planned to silently move in the dark to that station with the camera poised and ready.

I first went to the Tebrukunna Road area near Pioneer and found that rather than it being a quiet lonely road, it was wide, fast and busy with constant log truck traffic. I found a place to get my vehicle off the road and prepared for my first night. I set up my equipment, checked everything was working and waited. Evening came and there were bird calls, but no action on the ground. It was clearly not a good site.

Early next morning I packed up and moved back through Goshen and Goulds Country looking for a public phone to report my news home, then through Lottah and up to Poimena, which appeared to be a great site: a wide, grassed and mossy area surrounded by trees and no other humans nearby. I parked my vehicle in light scrub, located some well-used animal pads, set up my microphones and waited for darkness. At dusk the place came alive, with wallabies quietly grazing. At 4.00 a.m. I was woken by a loud crunching noise so I crept out of my vehicle and over to the particular bait station. I found the wire container torn open, bones and meat lying about, but no culprit. No doubt a Tasmanian devil had seen me approach in the moonlight and dispersed. On the second day I found my meat bait 'going off' so threw it all away and by the next morning it had all gone.

In this way I photographed many native cats (*Dasyurus viverrinus*), now known as eastern quolls, and watched them playing with one another while frogs sang in the distance. In the 1880s Poimena had boasted a large human population with tin mines, shops, a school and a hotel. I wandered around the deserted streets such as George Street, Campbell Street, William Street and Jerome Street, thinking of those early pioneer days.

I strolled up to Mount Maurice the next day for the great view and saw some old mine workings before leaving and travelling via Crystal Hill Road to Gladstone and located the road to Amber Hill. I managed to get the vehicle up a steep washed-out track to level ground among orchids, kangaroo tail (*Xanthorrhoea australis*) and she-oak (*Casuarina* spp.) and set up camp. During the night a noise started the recorder and woke me. I could see a small dark shape moving around the bait station, so I crept over, pressed the buttons on the camera and spotlight, only to see a cat.

The following morning, I walked up to see the Banca Dam before driving back to Pioneer and on to Legerwood to see my relatives, who convinced me that some of my recordings of crunching noises from Poimena were of 'something large' so the next day I obtained more pieces of meat from a local butcher and headed back, only to find the last section of road blocked by a large tree. I then moved to the hills south of Ringarooma, up Mathinna Plains Road and turned west toward Diddleum Plains. Once more I set up my gear just as wild weather was approaching. So, with strong winds and rain overnight that's all that I was able to record. I was very concerned that a tree might blow down across my van. The following day I returned home.

Summing up, I made some good daytime sound recordings of animals in the north-east forests. I tried to hide in the vehicle, lighting no campfires and trying not to release smells from cooking. I reported my location daily by amateur radio. The wildlife at Poimena was prolific, so I think it is an ideal location for a tiger to exist. However, my trip was a learning experience. I knew that a tiger would not necessarily be interested in carrion. I had difficulty sleeping during the daytime in order to stay alert all night and moving silently through the bush in total darkness was almost impossible.

Around this time another search was underway in the central highlands near Lake Adelaide with building materials being flown in to erect a large hut, using power equipment. A totally wrong approach in my opinion, as any tigers would run for miles from this noisy intrusion. I hope that one day a tiger will be positively sighted and then the species left in peace.

Why I think the Tasmanian wolf is still extant

Richard Freeman

As a cryptozoologist working for the UK-based Centre for Fortean Zoology (the world's largest cryptozoological organisation) I have spent over 20 years searching for creatures unknown to science or deemed, perhaps wrongly, to be extinct. I have hunted for the Mongolian death worm, the yeti, the orang-pendek, the giant anaconda, the almasty and many others, but the one that I am most convinced about is the continued survival of the thylacine or Tasmanian wolf. The Centre for Fortean Zoology chose this creature as its logo and emblem because it is an icon for both cryptozoology and conservation.

Unlike most other cryptids, we know that the thylacine existed and its date of supposed extinction is fairly recent, unlike *Gigantopithecus* (a giant Asian ape associated with the yeti), which died out 100 000 years ago.

The idea of a population of wolf-sized predators living undetected on an island since the mid-1930s may seem far-fetched, but although Tasmania is roughly the size of Ireland, it has a population of less than 541 965, most of whom live in either Hobart or Launceston. Towns in Tasmania are more like large villages and villages like hamlets.

Before I travelled to Tasmania, I did not realise just how wild it is. Now, after three visits I can fully attest that there is enough wilderness in the west of the island to hide a population of thylacines. There are many kilometres of untouched forest, where nobody lives and very few visit. During my visits, conducted with among others folklorist and thylacine researcher Mike Williams, I have spoken to many eyewitnesses who have no vested interest in the subject. Most do not want me to reveal the areas where their sightings occurred, for fear of the animals being hunted or disturbed.

As well as ordinary members of the public, I have spoken to a government-licensed shooter, contracted to kill feral cats in the bush. He had seen thylacines on two occasions. Another witness was an ex-logger, who had spent most of his life in the wilderness. Both his wife and his son had seen thylacines on separate occasions. Some sightings were multi-witness, including an instance when two car loads of people, who were not known to each other, saw a thylacine running along a road side. Another case involved a car full of workers on a hydro-electric scheme who encountered a thylacine in the road en route to work.

On one expedition, I was in remote forest in the east of Tasmania. On a section of the track, I became aware of an unusual smell, somewhat like that produced by a hyena. I am a former zookeeper, and a regular zoo visitor, so I am familiar with the smell. The odour seemed to intersect the track and was only noticeable in one area. It was as if whatever had left the scent had recently crossed the track. The thylacine was said to smell very like a hyena. Did I smell a thylacine? I don't know for sure, but there had been sightings in the area in the past.

Eyewitness accounts and suitable habitat aside, there are other factors that support the thylacine's survival into the present day. The creature's continued survival has been predicted by a computer program. Professor Henry Nix of the Australian University's Centre for Resource and Environmental Studies developed a computer program called BIOCLIM as a research tool. It matches what is known about a species' habits, preferences and geographical distribution and predicts where within a given area the target species would most likely be found. Nix applied his software to the thylacine and there was an almost perfect match between the areas where the program predicted a surviving population should be expected and the areas where sightings were being made.

A good example of how a population of large carnivores can remain undetected on an island is that of the Zanzibar leopard (*Panthera pardus pardus*). This creature was thought extinct since the 1990s, but was rediscovered on the island by camera-trap in 2018. Zanzibar has a land area of 2 463 km^2 and a population of 1 503 569. Tasmania in comparison has a land area of 90 758 km^2 but only a population of 541 965. The case hardly needs stating that if a predator can exist on the former it can easily exist undetected on the latter.

There are many iconic extinct animals, such as the dodo (*Raphus cucullatus*), the great auk (*Pinguinus impennis)* and the passenger pigeon (*Ectopistes migratorius*), that nobody reports seeing, yet people report the thylacine on a regular basis. Furthermore, the south-west of Tasmania was never settled, save for a handful of tin miners and fishermen at Port Davey. The area itself produced no thylacines during the bounty period and is not an ideal habitat for the thylacine, but we know that species under pressure can retreat to, and indeed thrive in, less than perfect conditions. A good example is the recently discovered population of Bengal tigers *(Panthera tigris tigris)* living in the high Himalayan mountains in Bhutan at an altitude of up to 3505 m (11 500 feet), far above their normal range. Therefore, it is quite possible that thylacine populations moved into the south-west during the bounty years and remained unmolested. Eventually these would have recolonised other areas of the island.

Today, most reports come from the north-east, west and west-coast regions of Tasmania. If I were a betting man, I'd put good money on the survival of the thylacine. I think it is just a matter of time until definitive proof of the species' continued existence comes to light.

Maybe, by the time you are reading this book, Tasmania's most magnificent animal will have re-emerged from the shadows into official existence once again. Interestingly, shortly after I returned from my first search for the thylacine I came upon a very interesting book with a striking passage about the Tasmanian wolf. *The Sea Inside*, written by Philip Hoare (2014), consisting of a number of essays on the world's oceans and certain islands. In the chapter on Tasmania, the author writes extensively on the thylacine and modern-day sightings, finishing his chapter with these words:

> What I do know is that in one institution I visit, a curator lets slip a quickly retracted remark, telling me it is not their secret to reveal. It is clear from what this person says, or does not say, that this strange half-life limbo of an animal which may or may not exist may soon be resolved, in its favour. That history is about to be reversed. That the thylacine is no longer extinct. If it ever was.

Thylacine eyewitnesses: the psychology of sightings

Michelle Vickers

Eyewitness testimony is very convincing. The more convinced the eyewitness is, the more convincing is the testimony. The problem with evidence from a witness is that the witness is human, and human perception and memory are prone to error.

It is unfortunate that while seeing might lead to believing, seeing itself may not be the truth. If we had certainty over each time a thylacine had been seen, it would no longer be considered extinct. Yet the most scientific of researchers have been unable to directly sight or validate the sightings of eyewitnesses over the past few decades, leading to controversy surrounding an eyewitness's claim.

Eyewitness sightings may be distorted without the witness being aware of this. Memory, unfortunately, is not a video camera. We tend to recall it like a film, but we are vulnerable to bias, to overly focusing on some details and missing others, and to exaggerate and minimise different components of what we think we have seen, all without our intention or knowledge.

To understand how eyewitnesses can believe they have seen a thylacine when they may not have, we need only look at the extensive psychological research investigating eyewitness testimony related to crimes. Similar to the witnesses of thylacine sightings, witnesses of crimes usually have only fleeting moments, maybe a minute or so, during which they need to see the person of interest, understand what they are seeing and commit to memory the features of the person they have seen. Surely if the eyewitness saw the criminal in action, they could describe them and recognise them? Unfortunately, no, the research, and actual court cases, have demonstrated this is not always the case.

It is not the fault of the eyewitness. It is the way human brains work. Memory is susceptible to error because memories are reconstructed from fragments of information.

We tend to reconstruct a memory into existing representations of the world to make our memories coherent. However, the reconstruction is susceptible to a range of errors that can lead us to believe we have seen or experienced something we may not have.

Take for example, if we are driving with a companion at dusk in Tasmania. All through Tasmania we hear about the 'Tasmanian tiger', the thylacine, and see pictures at every little town and tourist site. Locals and tourists alike are familiar with such images and are all well aware of the vast and protected Tasmanian wilderness where potentially the thylacine might to this day be safely avoiding human contact. It is possible and we are taught through these representations what one might expect to see if we are lucky enough to be a witness. While driving we see an animal run across the road and into the scrub. We both see the size, we see the colour, we see the stripes; we are in Tasmania after all – what else could it possibly be? We both check if the other saw the same thing. A slight smile or nod from the other and we are convinced.

Convincing indeed. Let's look at how our eyewitnesses may have this sighting wrong.

Firstly, there is the estimate that approximately 90% of the information that reaches the eye is lost by the time it reaches the brain (Gregory 1970), leaving fragments of what has been seen. Indeed, our witnesses may store the 10% of information that includes seeing a tan-coloured animal near the Tasmanian wilderness. Flickers of light from the fading sun and headlights of the car, reflecting off the foliage, may have added a memory fragment of stripes.

The reconstruction of the memory, which happened immediately, would then be influenced by existing representations from having seen many thylacine images, and the influence of the location builds a heightened expectation or desire to solve the mystery, to see the unknown. It's like when we walk into a supposedly haunted house and half expect to see the ghost. If our witnesses were not in Tasmania, there would be no consideration of having seen a Tasmanian tiger. It is an *expectation bias*.

The reconstruction of the memory may have projected the stripy foliage shadows onto the animal, causing an inaccurate mental image, an effect known as *illusory correlation bias*. Focusing on the colour and stripes, within the context of the location, is likely to cause what is known as *attentional bias*, whereby our witnesses' brains will be focusing on certain elements while not recalling details that may not match with the memory being reconstructed. In this case, if the animal was in fact a wallaby, a common tan animal of similar size found in Tasmania, our witnesses may have unwittingly not paid attention to the conflicting details.

When an eyewitness first tells of their memory, the response of the person they tell has been shown to affect the strength of their memory. Our witnesses turned to each other and had a leading question, suggestible to a positive answer. After receiving a positive response from each other, their own memories gain strength; that is, their belief in what they thought they saw moved away from having an element of doubt or curiosity to become their truth. This is known as *confirmation bias* and it is a common human trait. In addition, as humans we often fall into the trap of a stronger belief in what we think we've seen when the outcome of our witness statement would be desirable, such as is the case for eyewitnesses of thylacine sightings where the outcome would be proof of life!

This is how we can have two credible witnesses together – unwittingly – recalling a false memory of a thylacine sighted at dusk on a road in Tasmania. This shows how recent potential sightings can be controversial and require secondary evidence to allow a confirmation, beyond relying on the fallible psychology of humans.

Is the thylacine extinct?

Stephen R. Sleightholme

I have frequently been asked if I believe the thylacine is now extinct. Logically, after 85 years since the last known specimen died at the Hobart Zoo, the answer would be 'yes'. However, I am constantly reminded of Eric Guiler's (1985) warning:

> It never ceases to surprise me that since 1936 it has been lamely accepted that the Thylacine was extinct or nearly so, even in the face of persistent sighting reports, some of which will stand considerable critical examination. This is a Tasmanian tragedy and it is disappointing that no world fauna body has sponsored a thorough search for them, the rarest of the world's mammals.

Was Guiler correct in his assessment that the extinction event might not yet have taken place? Many would argue that he was; others that he was clinging to false hopes. Yet Guiler, a renowned authority on the species, was working on a field search in the far north-west of the state in 2002 when he suffered his debilitating stroke. So, even in the face of what would seem to be formidable odds, he never gave up on the species.

Most authorities now accept that the death of the last captive specimen in 1936 was highly unlikely to be the extinction event. Sleightholme and Campbell (2016) state: 'These extinction dates have broadly been accepted by the scientific community as fact, despite logic dictating that a secretive nocturnal animal, concealed amongst vast tracts of extremely dense vegetation, would persist well beyond the last known capture on the fringe of its habitat'. Mooney (2014) concurs and goes as far as providing an estimate of the numbers remaining in the wild when the last wild capture was made: 'To me it is entirely probable that at the time the last Thylacine recorded caught or killed suffered this indignity in 1933, there were at least one hundred Thylacines left'. He continues: 'As to the fate of those last 100 thylacines remaining in 1933, my suggestion is they were fragmented into small groups – some of which just fizzled out like flags being brushed off a war game'. The species, with near certainty, survived beyond the 1930s.

From the 1940s to the present day, reports are regularly received, many from respected sources, of thylacines seen in the wild. Undoubtedly, many of these sightings will be misidentifications, but not all. Some are extremely difficult to easily dismiss, as they were made by park rangers or biologists, all of whom would be familiar with the species. The most publicised of these sightings was that made by Hans Naarding at Togari in 1982. Naarding, an experienced field ranger with the Parks and Wildlife Department, observed a thylacine at close quarters for several minutes. The sighting was kept confidential for 2 years to allow a proper investigation of the area to be carried out. Unfortunately, no evidence was found of the thylacine, but the official report into the sighting concluded: 'The search area was used irregularly by Thylacines up until autumn 1982 but use has diminished due to increased disturbance to the point that detection of animals in not probable, despite large efforts. Unless the Thylacine observed in March 1982 by the service biologist was the last of the species it must be accepted that Thylacines survive in a number of areas of Tasmania'. So, should we be cautiously optimistic that the species is still clinging on in a remote part of the state?

Over the past century, a number of species that had previously been declared extinct have been rediscovered in the wild. These are often referred to as 'Lazarus' species. The mahogany glider (*Petaurus gracilis*) declared extinct in 1886 and rediscovered in 1989; the takahe (*Porphyrio hochstetteri*) declared extinct in 1898 and rediscovered in 1948; the Bermuda petrel (*Pterodroma cahow*) declared extinct in 1620 and rediscovered in 1951; the brindled nail-tailed wallaby (*Onychogalea fraenata*) declared extinct in 1930 and rediscovered in 1973. Gilbert's potoroo (*Potorous gilbertii*) declared extinct in 1874 and rediscovered in 1994 and Leadbeater's possum (*Gymnobelideus leadbeateri*) declared extinct in 1909 and rediscovered in 1961. Like the thylacine, many of these species are timid by nature, nocturnal by habit and forest edge or forest dwellers, so are naturally adept at avoiding detection.

In an effort to resolve the survival quandary, Brook *et al.* (2021) conducted what was to be the largest survey of sighting reports ever undertaken. The team, of which I was privileged to be part, examined over 1200 reported sightings from 1910 to the present day and concluded:

> In sum, this collective body of evidence and associated analyses indicates that the continued persistence of the Thylacine is unlikely – but possible – and that the true extinction year, if the species is indeed now extinct, occurred much later than the commonly held date of 1936. Indeed, the inferred extinction window is wide and relatively recent, spanning from the 1980s to the present day, with extinction most likely in the late 1990s or early 2000s. While improbable, these aggregate data and modelling suggest some chance of ongoing persistence in the remote wilderness of the island.

So, where does this leave us with the survival/extinction debate? Although good-quality sighting reports can be considered positive indicators of survival, Paddle (2000) justifiably states: 'The criterion for establishing the existence of the thylacine beyond 1936 can only be met through the production of a body, either dead, or, preferably, alive', and to date, this has not been met. Therefore, the jury remains out on the question of extinction and we are left pondering an unresolved quandary, with numerous credible sighting reports, but no physical body for verification. As Campbell (2021) poignantly notes: 'The thylacine was always elusive, perhaps that is still the case'.

A second extinction: was a host-specific parasite lost too?

Liana F. Wait

If thylacines hosted their own unique parasites, it is quite likely that we lost more than one species when the species was driven to extinction.

As well as being agents of disease and death, parasites also play important roles in ecological communities and ecosystems, where they function as 'ecosystem engineers': parasites help shape host population dynamics, balance competition between free-living species and provide important links within food webs (Carlson *et al.* 2020). Parasite conservationists describe them as being the lynch pins holding ecosystems together – if we remove too many of them, the structural integrity of the whole system will collapse. However, parasites are particularly vulnerable to extinction because they intrinsically rely on their hosts: if the hosts become scarce or even extinct, the parasites that call them home are doomed. In some cases, parasite extinction actually precedes host extinction, because most parasites require the population density of their hosts to be above a certain threshold in order for transmission to occur.

Although some parasites only infect one type of host for their entire lives, many parasites have complex life cycles, with different stages living in different hosts or environments – similar to the different life stages of a butterfly, though maybe not quite so photogenic. For example, most tapeworms live inside herbivorous animals during their larval stage, whereas the adult tapeworm stage lives and reproduces inside carnivorous animals that become infected when they eat the flesh of infected herbivores. The life cycle is completed when uninfected herbivores ingest tapeworm eggs shed in the faeces of infected carnivores – the circle of life!

This is a generic life cycle, but most tapeworms are quite specific about which particular carnivorous host they will live and reproduce within, though less fussy in terms of which species their larvae will infect. For example, adult *Dasyurotaenia robusta* tapeworms have only ever been found in the intestines of the Tasmanian devil and these tapeworms are one of very few parasites to be listed as a target for conservation (Beveridge and Spratt 2015). But because this parasite also infects other hosts during other parts of the life cycle (e.g. the herbivores in the generic example above), the removal of one of these hosts effectively prevents completion of the lifecycle.

We know very little about the parasites of the thylacine. Only two species of parasitic worm (one nematode and one tapeworm, neither of which are specific to the thylacine (Sprent 1972; Obendorf and Smith 1989)) and one species of flea (also found on several other marsupial species, including the Tasmanian devil (Dunnet and Mardon 1974)) are recorded and there are no reports of viruses, bacteria or protozoa from the thylacine, though they undoubtedly hosted parasites in all of these classes. There is ongoing controversy over the role of disease in the decline and extinction of the thylacine (Paddle 2000, 2012; Prowse *et al.* 2013; Sleightholme and Campbell 2016), but the hypothetical causative agent of this largely anecdotal disease is unknown and, in any case, it is claimed that this disease also infected Tasmanian devils (Paddle 2000, 2012), and so it likely did not go 'co-extinct' with the thylacine.

As one of the thylacine's closest extant relatives (both phylogenetically, and in terms of its functional role as a carnivore within the Tasmanian ecosystem), we might try to estimate how many parasites infected the thylacine, based on how many parasites infect Tasmanian devils. Tasmanian devils have been reported to be infected by 28 different parasite species: 5 are Tasmanian devil-specific, or have Tasmanian-devil specific portions of their life cycle, 16 others infect multiple marsupial hosts, and 7 appear to have originated from introduced host species (Wait *et al.* 2017). Some studies suggest that the number of different parasite species harboured by a given host species increases with host size (Kamiya *et al.* 2014), so we might predict that thylacines hosted even more parasites than Tasmanian devils. Little is concretely known about the behaviour and diet of the thylacine, but these factors could also have played a role in determining how many different parasites infected them. Social behaviour increases the transmission of some parasites and so if thylacines were social animals, we would predict that they harboured a higher number of parasite species, and if thylacines preyed upon different (larger) prey species than other carnivorous marsupials, then it's also more likely that they harboured their own unique parasites, compared to the scenario where their diet completely overlapped with those of Tasmanian devils and quolls.

It seems likely that at least one parasite species went extinct alongside the thylacine. Although this might not seem like much of a tragedy, such a loss could have had important implications for Tasmanian ecosystems. Think about parasites the next time you think about conservation – they might not be charismatic, but they are important team players in maintaining and balancing our world's ecosystems.

Thylacine habitat increases after the British invasion of Tasmania

Jamie B. Kirkpatrick

The habitat of an organism needs to be physically capable of providing the organism with both sustenance and shelter to a degree that allows successful reproduction through the generations. Organisms do not necessarily occupy all of their potential habitat. They may not be able to get there, or other species may prevent them from occupying it by various forms of interference, including predation, density-dependent disease and competition for resources. Thus, the availability of habitat is just one influence on the distribution and abundance of individual species.

To understand the habitat of the thylacine we need to understand the habitat of its prey. The morphology of the thylacine suggests a less than 1–5 kg prey size range (see pp. 20–3) captured by ambush and pursuit (see pp. 17–18). However, thylacines occurred in alpine country where small mammals are rare and wallabies and wombats abundant, and as there is some evidence of the species hunting in small groups (Paddle 2000), it may be best assumed that the species predated most native mammals, at least as juveniles. Lambs may also have been prey, as may have the rabbits, hares, cats, rats and mice associated with land modified by the British yeomen who displaced the Palawa from their land.

The best original habitat for thylacines, native grassland and grassy woodland (Sleightholme and Campbell 2016), has suffered substantial depletion in Tasmania since the British invasion, with only 35% of the original 500 000 hectares extant in the 21st century (Mendel and Kirkpatrick 2002). Most of the remaining areas of native grassland and grassy woodland are at high altitude or in hilly country, with lowland flats being almost entirely converted to exotic cover.

The availability of prey for a non-arboreal predator such as the thylacine is related to the primary productivity at ground level of the plants, fungi and invertebrates that they consume. In Tasmania, high productivity in the foodstuffs of such prey is associated with a well-insolated (sun exposed) vegetation ground layer on basic igneous rocks, alluvial deposits and sedimentary rocks rich in calcium carbonate. However, thylacines were recorded in most environments in Tasmania, albeit at a lesser incidence than in the grassy country. Apart from the coast where soils are enriched by salt spray and marine animal strandings, the quartzite country of south-west Tasmania is very poor thylacine habitat.

Grasses and forbs best support macropods and wombats, whereas fungi and invertebrates support potoroos, bettongs and bandicoots. The most fertile country dominated by grasses and herbs is agricultural land. When stock are taken off, as on Maria Island (Ingram and Kirkpatrick 2013), or native animal control ceases, pastoral land supports dense populations of native mammals. Approximately 2 000 000 hectares of Tasmania are now pastoral and agricultural land dominated by exotic plants. This land could be prime habitat for thylacines if they were not killed by landowners, as happened in the 19th century (Kirkpatrick and Bridle 2007).

The cessation of both snaring and widespread 1080 poisoning campaigns, and legal protection, indicate a potential for an expansion of thylacines from remote refugia into the settled parts of Tasmania. Here more thylacines could possibly be supported than were present prior to 1803. As cryptic, crepuscular scavengers of flesh and killers of small game they might penetrate towns, suburbs and exurbs, like the Tasmanian devil, wallabies, pademelons and many smaller mammals (Daniels and Kirkpatrick 2012).

Google Trends data for thylacine-related keywords (2004–20)

Michael Zieger, Steffen Springer and Branden Holmes

The Tasmanian tiger (*Thylacinus cynocephalus*), commonly known as the thylacine, is understandably still of great interest to people around the world in the 21st century. Evidence of this can be found by evaluating search queries directed at internet search engines. A suitable (and popular) tool for this purpose is Google Trends, which was introduced to map the interests of internet users on a regional and global level. Google Trends data are available from 2004 to the present and can be used, for example, to study environmental and nature conservation issues (Nghiem *et al.* 2016; Soriano-Redondo *et al.* 2017; Zieger and Springer 2021).

Because up to five search terms (keywords) can be queried at the same time, with the relative search interest shown both over time and according to regional distribution, we used Google Trends data to examine the relative worldwide popularity of four English-language thylacine-related search terms such as 'thylacine' and 'Tasmanian tiger' between 2004 and 2020 (Plate 17). Compared with the relative search interest for the largest surviving carnivorous marsupial, the Tasmanian devil (*Sarcophilus harrisii*), the search interest in thylacine-related keywords has consistently remained lower. Of the thylacine-related search terms studied, 'Tasmanian tiger' is the most commonly used worldwide, with the species' scientific name (*Thylacinus cynocephalus*) far behind the more popular terms.

Unsurprisingly, the relative regional interest in the term 'Tasmanian tiger' has a clear focus on Australia. If only Australian searches are considered, the focus here is on Tasmania, as already shown elsewhere for 'thylacine' and thylacine 'sightings' (Zieger and Springer 2020).

Hardly any seasonality could be found for the search term 'thylacine sightings' for search queries in Australia, although searches for 'National Threatened Species Day' and 'Tasmanian devil' showed a certain seasonality (Zieger and Springer 2020).

Therefore, Google Trends may offer a tool to assess seasonal patterns in wildlife (Zieger and Springer 2020). In addition, relevant regional search queries can also be documented with Google Trends in order to prove and investigate the interest of the population. However, some limitations of Google Trends also need to be considered, especially with low search volumes (Dietzel 2016). Although the specific search motivations of users cannot be clearly assigned, analysis of the Google Trends data nevertheless provides an important source of information to illustrate the interest in certain topics among the human population at the regional and global level.

There is still a major human interest in the two iconic species, *Thylacinus cynocephalus* and *Sarcophilus harrisii* – both worldwide and, in particular, Australia. The interest in National Threatened Species Day, for example, but also in devil facial tumour disease, may show that, in addition to current internet user interests, there is also a sense of responsibility among the human population for nature conservation and environmental protection (Zieger and Springer 2020). Further raising of public awareness of the current biodiversity crisis may therefore help prevent losses such as that of the unique thylacine.

In the shadow of the thylacine

Kenny J. Travouillon

The thylacine is by far the most well-known extinct Australian mammal, with National Threatened Species Day commemorating each year the anniversary of the death of the last thylacine at Beaumaris Zoo on 7 September 1936. But what about other extinct mammals? On the official National Threatened Species Day website,[5] they do not even get a mention.

There are currently 34 mammal species listed as extinct in Australia, by far the worst extinction record of any country in the world (Woinarski *et al.* 2019), and the number is likely to go up with climate change and new scientific discoveries. Climate change has already claimed its first victim, the Bramble Cay melomys (*Melomys rubicola*), a native rodent found only on a small island in the Torres Strait. As the sea level raised, the island was completely submerged, and the Bramble Cay melomys perished (Waller *et al.* 2017). Museum collections are a source of extinct species discoveries, with new species being discovered recently (e.g. Nullarbor barred bandicoot, Yirratji) from specimens collected in the early 1900s before they went extinct (Travouillon and Phillips 2018; Travouillon *et al.* 2019).

The Bramble Cay melomys is not the only island species to have gone extinct. The dusky flying-fox (*Pteropus brunneus*) has been lost on Percy Island in Queensland, the Lord Howe long-eared bat (*Nyctophilus howensis*) on Lord Howe Island and, worst of all, Christmas Island has lost two rats (Maclear's rat, *Rattus macleari*, and bulldog rat, *Rattus nativitatis*), the Christmas Island pipistrelle (*Pipistrellus murrayi*) and the Christmas Island shrew (*Crocidura trichura*). On the Australian continent, the casualties can be classified into three mains groups: bandicoots and bilbies (Peramelemorphia: 8 species lost), rodents (Rodentia: 10 species lost) and bettongs, rat-kangaroos and wallabies (Macropodiformes: 8 species lost). The thylacine is the only member of Dasyuromorphia to have gone extinct.

Of all the extinct mammals in Australia, the thylacine is probably the best represented in museum collections, with hundreds of specimens scattered around the world. This is not the case for the other 33 species. Most are represented by perhaps a dozen specimens, but some by a single specimen. This makes it very difficult to reconstruct the past distribution of extinct species and recover any information about their ecology. Subfossils have played an important part in documenting their past distributions, as well as early notes from 19th century naturalists. By far the most valuable information came from Indigenous elders who remembered seeing or hunting some of these species up until the 1960s. They remembered some of their ecological habits, what food they ate and how they nested (Burbidge *et al.* 1988).

The causes of extinction are varied. On Christmas Island, diseases brought by the black rat (*Rattus rattus*) are likely to have caused the extinction of the two endemic rats. The introduction of predators such as cats and foxes on the Australian mainland has been linked to the demise of most rodents, bandicoots and wallabies, especially species in the critical weight range between ~35 g and 5500 g (Burbidge and McKenzie 1989). Competition with introduced herbivores, such as rabbits, sheep and cattle, would also have played a significant role. Land clearing for the development of agriculture and changes to fire regimens removed valuable habitat and food and likely have affected the Liverpool Plains striped bandicoot (*Perameles fasciata*) in New South Wales and the marl (*Perameles myosuros*) in the wheatbelt

5 https://www.environment.nsw.gov.au/topics/animals-and-plants/threatened-species/saving-our-species-news/threatened-species-day

of Western Australia. The timing of their extinction seems to follow the combination of at least two factors, with rodents being affected first, mostly due to the early arrival of cats in the 1800s (Roycroft *et al.* 2021), and bandicoots and wallabies going next with the arrival of foxes and increased land use.

Apart from the recent extinctions of the Bramble Cay melomys and Christmas Island pipistrelle, it seems that we have managed to slowdown the processes that have caused the extinction of mammals in Australia, as there have not been any continental extinctions since the 1970s. However, we still have a huge number of mammals listed as endangered or vulnerable (Van Dyck and Strahan 2008) and efforts to conserve them should not by any means slow down. In our effort to conserve our remaining mammals, we should remember what species we have lost and make sure that we do not lose any more.

PART 8: BEYOND THE PRESENT

Analysing scat samples to learn about elusive animals

Catherine Grueber

As animals move through the landscape, they leave traces of themselves. A distinctive scent marking on a tree trunk, the grisly remains of a recent kill or perhaps more ordinarily, faeces (scat). Monitoring secretive or rare species by picking up scats has aided the conservation of jaguars in Brazil, chimpanzees in Gabon and the Eurasian lynx, among many others. Species that particularly benefit are those that are difficult to find, because scat studies enable conservation biologists to study the movements and health of individual animals they might never have seen. If a thylacine scat were found in the Tasmanian wilderness, it too would provide a range of valuable information, all without even setting eyes on the animal itself.

How would one know whether a particular scat on the forest floor was produced by a thylacine? We would first need to rule out other animals with similar body size, diet, anatomy or habits. For example, carnivorous species such as Tasmanian devils, quolls, cats, foxes and dogs might feasibly produce similar scats (is it a large devil scat or a small thylacine?). If thylacine were suspected, the next step would be to send the sample for genetic analysis as soon as possible after collection. Because scat samples are collected from the environment, there is a high potential for contamination and sample quality degrades rapidly due to exposure to light, moisture, microbes etc. Fresh samples yield the most reliable data and appropriate storage to avoid contamination and preserve the DNA (e.g. freezing or using a preservative) are essential (Panasci *et al.* 2011).

What can we learn from genetic analysis of scat samples? In animal digestive systems, as the gastrointestinal content moves through, a small number of intestinal epithelial cells are sloughed off and end up attached to the deposited scat. Just as with a blood or hair sample, these cells contain DNA that can provide a genetic profile of the individual. Decades of genetic research have produced reference profiles of many species, including those that are likely to leave scats in the Tasmanian forest. Provided the sample is of good quality, geneticists can carefully extract DNA to determine which species is most likely to have produced it. It may also be possible to use variations in the genetic code to determine whether the scat came from a male or female (Berry *et al.* 2007).

Scat samples don't only contain the DNA of the animal that produced them, they also contain trace amounts of the DNA that was within the plants and animals that were consumed. Although visual inspection of a scat might reveal fragments of feathers that have passed through the digestive tract, DNA analysis can help identify what type of bird had an unlucky encounter with this carnivore. This type of dietary information shows us how species interact within their environment, but requires very careful sample handling to avoid contamination (Goldberg *et al.* 2016; Galan *et al.* 2018). Furthermore, wild animal scats can be used to characterise the bacterial ecosystem of the gut, known as the microbiome. Many readers will be familiar with the importance of maintaining a healthy balance of gut bacteria: we could compare the microbiome of a thylacine scat to published references to learn more about that individual's health, as was done for coyotes in North America (Sugden *et al.* 2020).

The methods summarised above hint at the wealth of knowledge that could be gained from even a single, carefully preserved scat sample. This knowledge supports a range of conservation interventions to protect rare and threatened species. By learning about species from their scats, including their diet and health, biologists are conserving a range of threatened species such as the Australian sea lion, northern bettong and the Tasmanian devil (Grueber *et al.* 2020).

Using environmental DNA (eDNA) to find the thylacine

Mieke van der Heyde

If thylacines (Tasmanian tigers) are still wandering around Tasmania, environmental DNA may be the key to finding them. Animals can shed DNA through skin cells, scats, hair and saliva, which can then be detected and identified using molecular methods (Taberlet *et al.* 2012). One of the main reasons to use eDNA is the incredible sensitivity of the tool. Once the DNA has been extracted from an environmental sample, polymerase chain reaction (PCR) is used to make millions of copies. The PCR works like a molecular magnet that sticks to the DNA we want (thylacine) and amplifies it, while ignoring everything else. As a result, we can detect very small concentrations of DNA (Furlan *et al.* 2019).

Many eDNA surveys involve targeting a 'barcoding' gene region that can be used to identify certain animals (e.g. fish) and sequencing that region to compare it with a DNA database (Taberlet *et al.* 2012). Although this will give you data on many species in a single sample, rare species may get 'drowned out' by the more common animals (Plate 17). A species-specific PCR assay will ignore all non-thylacine DNA and detect only the presence or absence of thylacine, similar to the way a COVID-19 PCR test detects the SARS-CoV-2 (coronavirus). Not only is this method more sensitive, it is also cost-effective because there is no need to sequence the DNA. Instead, we use quantitative PCR (qPCR), which is cheaper than DNA barcoding, to tell us whether or not there was any thylacine DNA in the sample to copy (Plate 18). We are fortunate that there are existing reference samples for the thylacine that can be used to create such assays (Feigin *et al.* 2018).

Unfortunately, validating eDNA probes in the field is problematic for a cryptic species. The most common source of eDNA is water, so river samples are often used to detect not only aquatic diversity, such as fish and crustaceans (Cowart *et al.* 2018; West *et al.* 2020), but also terrestrial animals (Sales *et al.* 2020). This type of broad eDNA survey could be useful to identify areas to focus on for more intensive searches. However, for especially rare species the DNA in water may be too dilute to be reliably detected. There are similar concerns with soil, where DNA deposition is likely to be 'patchier' than in water where there is more mixing. An additional concern with soil is the possible preservation of DNA for decades or more in the right (cool and dry) conditions. In order to maximise the chances of detecting the thylacine we may need to target the sample types most likely to have had recent interaction with the animal.

One way to do this is by using other animals as DNA collectors. For examples, leeches can carry concentrated blood for several months, are common in terrestrial waters in Tasmania and are easily sampled because they readily attack humans (Schnell *et al.* 2015). As a result, not only is the DNA more concentrated, it is also preserved longer than in water, where DNA may only last a few hours or days (Barnes *et al.* 2014). However, the corresponding downside is that any positive detections may be months old and the animal may be long gone from the area. Although in the case of putatively extinct animals such as the thylacine, even this months-old DNA would effectively constitute the rediscovery of the species.

Another way to target samples more likely to contain thylacine DNA is to use eDNA to verify suspected traces, such as scats, tracks or hair. Scats, or faeces, are commonly used as evidence of animal activity and DNA can confirm suspected identifications (Wadley *et al.* 2013). In addition to scats, the ability to determine whether tracks belong to the thylacine or not would be a valuable tool in the search for this cryptid. Franklin *et al.* (2019) managed to detect rare carnivores (the fisher (weasel family), lynx and wolverine) from tracks in the snow using qPCR probes. This method would theoretically also work for tracks found in soil or mud, but the advantage of tracks in the snow is the guarantee that the tracks are recent (i.e. since the last snowfall) and thus more likely to contain mostly the DNA of the animal making the tracks. Finally, a common method used to monitor rare carnivores is hair traps. These simple devices, which may be baited, will snag samples of fur from an animal as it walks by (Castro-Arellano *et al.* 2008). Strictly speaking, this sample type may not be considered as eDNA by some, as it contains pieces of the animal rather than just DNA fragments. However, it is included in this chapter because the methods used to verify whether or not the fur belongs to the thylacine are exactly the same in the laboratory.

Overall, eDNA is a powerful, cost-efficient tool with incredible sensitivity that can be used for broad surveys or to verify suspected traces. The search for the thylacine will likely require multiple lines of evidence and the best way to use this tool to detect cryptic species is uncertain. However, the potential is clear and eDNA may prove the key to unlocking the mysteries of the thylacine.

Using technology in the pursuit of evidence

Mike Williams and Rebecca Lang

Researchers of rare and endangered animals have been making use of cameras in the field in Australia for more than 70 years. And while there have been some successes – recent photographs of the night parrot (*Pezoporus occidentalis*) come to mind – no-one has yet procured a clear, indisputable image of the thylacine. Technology has also improved significantly during that time, so for the amateur researcher working on the premise that a population of thylacines might still exist, searching for evidence using modern technology should in theory advance the cause.

Game cameras

In the past decade, 'game cameras', small portable cameras powered by batteries and loaded with memory cards, have become so common that you can purchase them for as little as A$100 on ebay or the local Aldi store. There are many models to choose from (we have used Acorn and Bushnell, to name a couple), but all game cameras are activated by a passive infrared motion sensing system, which picks up on the movement of a living animal with a body temperature different to that of the ambient air near the camera.

However, there are plenty of other things that can trigger the sensors of the camera, including a low battery, a sudden alteration in temperature and the movement of nearby vegetation. There is also some discussion around whether animals may be deterred from stepping in front of cameras because they can sense the system due to sounds emitted by certain units while in standby mode, though a more likely explanation for this apparent reticence is the human scent on or around the unit. However, if these units are used carefully, there is not an animal on the planet that cannot be captured by them.

Yet even in optimal circumstances all may not be what it first seems to be: several years ago during an expedition to Tasmania, road-kill wallabies were arranged in front of a game camera. Upon returning hours later, 'something' had very cleanly eaten out large sections of the carcasses. Because whatever had eaten the carcasses had done so very precisely, and having observed that a devil's feeding habits are anything but precise, we were certain it must be a thylacine. Yet, when we played the footage back, it clearly showed devils at the carcasses. Apparently, some devils have better table manners than others. On another occasion, we were fascinated to find a goshawk had also been busy dismantling a carcass in front of the camera, as well as numerous feral cats (always black), some curious rats and quolls, but no thylacines.

Drones

Drones are perhaps some of the most exciting technologies to be made available in recent years, but that availability belies the skill level and additional 'bells and whistles' required to realise their true potential. For an animal believed to be primarily diurnal or nocturnal, the potential low-light filming capabilities required to potentially capture a thylacine image presents a significant obstacle. This is where thermal video with real-time radio frequency and 'back to base' station receivers are required. Likely, if the drone does film a creature in low-light conditions, it will be of the 'thermal quadrupedal blob' variety. In other words, unclear or unidentifiable. There is a price to be paid by muliplying your gadgets; due to the additional weight of the thermal camera you are likely left with a battery airtime of about ≤20 min.

This is all supposing your expensive drone avoids colliding with anything, fixed or moving, and doesn't become lost in a thickly vegetated or inaccessible area. Drones can reach places you may never ordinarily be able to visit, but they also represent a significant and sometimes risky investment, with an uncertain return. A quality drone will cost A$1000–2000 (there are cheaper models but they fall more into the 'toy' category) and a top-quality drone will leave you ~A$10 000 out of pocket.

Dash cams

Many thylacine sightings have been made from moving cars, often at night, while the witnesses were driving to or from their home or place of work. So, what better equipment could there be than dashboard-mounted camera, or dash cam. Although this might, initially, seem like a good idea, such equipment has its limitations. Most models are very poor at capturing images at night unless you are within close proximity to your target, the camera has very little focal length between the lens and the sensor. Practically, the effectiveness of dash cams is determined by their resolution and field of view. A minimum class 10 card is required for storage.

Most models can be set to loop after a determined amount of time and will automatically record over old footage unless it is marked and saved. Therefore, if you *do* happen to encounter a thylacine crossing the road, it is important to remember to get that 'footage of the century' off your dash cam and, in the heat of the moment, remember to switch it off or mark and save the footage. If you are really keen, and technically minded, you could also mount a full-frame camera to your dashboard using a platform capable of accommodating the vibrations of your car and of the road on which it is travelling. In ideal circumstances, a good-quality dash cam will capture serviceable footage of an animal, *if* it is within 20 m of the car during daylight hours. As with other pieces of equipment you get what you pay for: a simple entry-level camera may cost as little as A$50–100, soaring to as much as A$1200 for one with a wider field of view and greater storage capacity.

Parabolic microphone

The sounds the thylacine supposedly makes vary between 'coughs', yaps/yips, barks, hisses and even alleged wolf-like howls. There are no archival recordings available for reference and only a few first-hand descriptions from witnesses, zoo workers and naturalists.

Guiler (1958) described the vocalisations of thylacines while hunting as: 'a coughing, barking noise, a low growl when irritated (probably a warning) and a whining noise like that of a puppy. The latter sound is perhaps a method of communication between individuals'. How then to capture these much sought-after vocalisations? Because the thylacine may be at a distance, you would need something that could collect and focus sound waves. Enter the parabolic microphone with its curved dish and centrally mounted microphone.

Prices vary between as little as A$80 to ~A$1200. The best equipment seems to be aimed at the 'twitching' fraternity, your friendly neighbourhood birdwatchers.

Body camera/Go Pro

In the past few years body camera prices have dropped significantly, making them accessible to many sporting and hobby enthusiasts, making the old excuse 'I never had time to reach for my camera' redundant. The models used by the authors cost around A$150–250. We have never tested the battery to its maximum charge time, but it is claimed to last up to

8 h. Though the cheaper models possess a minimal focal length, the footage they provide is reasonably good. Though more expensive, the Go Pro action cameras offer an impressive, high-quality way to capture whatever is happening wherever you are, even if the thylacines still avoid your lens.

Conclusion

There are also more expensive technologies such as starlight cameras and thermal cameras, but these are invariably out of the price range of most thylacine seekers. Digital cameras and camera phones are mundane enough to exclude from this brief account. The advent and application of modern technology in the search for the thylacine, while promising, is not necessarily a golden ticket to proving its continuing existence. Lastly, spare a thought for those who have claimed over the years to have captured that 'footage of the century', only to mislay it, like camper John Zaundevelle who, along with fellow members of the Victorian Bushwalking Teams and Search Party Association, claimed to have encountered, and photographed, several thylacines near their Lune River campsite in south-eastern Tasmania in 1986.

> They said they followed the animals for some time, sighting them clearly at least three times and seeing them go into an underground lair. They said there had been one larger, older animal and two younger animals, possibly cubs. A party member, Mr John Zaundevelle, is said to have a series of sketches of the animals and of paw prints in sand in the area.
>
> Mr Zaundevelle could not be contacted yesterday and was believed to have returned to the area to retrieve a camera that had become wedged in a rock crevice after one of the party members fell. The camera is said to contain a film with seven shots of the animals (*The Canberra Times* 1986).

No subsequent mention has been found as towhether or not Mr Zaundevelle ever recovered his precious camera with its indisputable evidence. Perhaps it was eaten by devils.

The thylacine: wanted dead, or alive?

Peter B. Banks and Dieter F. Hochuli

> Sometimes dead is better.
>
> – tagline from *Pet Sematary*, Paramount Pictures (1989)

On 23 February 2021, the world woke to news that naturalists had obtained convincing new footage of a thylacine (*Thylacinus cynocephalus*) in the wild. This was nothing new – reports of potential thylacine sightings appear regularly enough, especially in the world of cryptozoology. But the thylacine is not the typical target sought by cryptozoologists hunting 'species' unknown to science, such as Nessie or Bigfoot. We very much know the thylacine: the species is officially considered extinct since the last individual died in the Beaumaris Zoo in 1936. What do they hope to achieve by finding it alive?

In all the excitement of its potential rediscovery, it is worth taking the time to wonder about the consequences if one time a sighting turns out to be real? Perhaps one time the blurry image won't be of another meandering canine, a pademelon tail or wallaby rear end, to show the thylacine isn't extinct after all? It would, of course, be wonderful for thylacines as a species and lead to all sorts of efforts to conserve it. But what would its rediscovery mean for conservation in general?

There is little doubt that the thylacine is an icon of extinction, not just in Australia but globally. Its demise is emblematic of the human impacts on wildlife that drive our current biodiversity crisis. It was essentially hunted to extinction because of money – either the money allegedly lost when thylacines killed livestock or the money made from the bounty on their heads. And while humans didn't pull the trigger to kill the last animal (unlike in the movie *The Hunter*), the thylacine was hunted to such low numbers, and its environment changed so much, it was sucked into an extinction vortex, scuppering the belated last ditch efforts to protect it. It is a sad story of extinction that has been retold too many times for too many species. Although deliberately driving species to extinction is almost never humanity's goal, we've created the circumstances that have meant too many species just faded away with no hope of recovery.

As such, the thylacine is a potent symbol of our extinction guilt. We have mourned its loss for almost 90 years (van Dooren and Rose 2017), even though in the same time we have seen dozens of other species go extinct in Australia. Most people do not know we lost a native mammal species just 3 years ago when the Bramble Cay melomys (*Melomys rubicola*) was declared extinct, the first mammal species lost forever due to climate change (Woinarski *et al.* 2017). Fewer still know we lost the Christmas Island pipistrelle (*Pipistrellus murrayi*) just a few years earlier, despite alarm bells being rung by conservation biologists for 20 years.

But there's a good chance everyone in Australia knows the plight of the thylacine and it motivates them to support conservation efforts to prevent more extinctions. We commemorate Threatened Species Day in Australia on 7 September, the date on which the last known thylacine died. National Geographic uses its story to educate kids about extinction. Indeed, the thylacine's story is being used in this very book to raise money to protect the Tasmanian devil from extinction. What then if this latest sighting is true, that thylacines have been out there in the Tasmanian wilderness all along?

In many ways, rediscovering the thylacine is akin to the efforts to recreate the thylacine in the lab. Enthusiasm for de-extinction of lost species remains a contentious issue for ecologists (Kohl 2017; Swart 2018). Although novel and technologically ambitious, the entrepreneurial proponents of de-extinction also promote one dangerous argument: that extinction isn't forever (Banks and Hochuli 2017).

Will rediscovering the thylacine mean the biodiversity crisis isn't that bad after all? Not at all, but certainly it would make some people question whether other reports of extinctions are really true. There is a very good chance its rediscovery would almost certainly be used by those with vested interests to undermine conservation efforts. Its iconic status would likely be used to argue that conservation has been 'crying wolf' all these years (Ladle *et al.* 2004), that the 'extinction crisis' is fake news being used to line the pockets of environmental grifters. Just like a cold snap in summer is used to argue global warming is a hoax (Fischer and Knutti 2014), the rediscovery of the thylacine will inevitably be used to argue the biodiversity crisis is another hoax and we just need to look harder for all these 'so-called extinct species' (Fisher and Blomberg 2011).

We have argued elsewhere that the thylacine is a powerful martyr for the conservation cause (Banks and Hochuli 2017). In being extinct, it arguably does far more for conservation efforts than it could if resurrected via de-extinction or rediscovered by cryptozoologists. In many ways, conservation needs its icons of extinction, species such as the dodo (*Raphus cucullatus*), the passenger pigeon (*Ectopistes migratorius*) and the thylacine, to stay extinct. As for all these flagship species for extinction, each time a thylacine sighting turns out to be false, we are reminded that for almost all species extinctions will indeed be forever. And as a consequence, extinction is something we must do everything to prevent for species still clinging to their existence.

Epilogue

Less than 24 h after the news broke of footage showing the thylacine had been rediscovered, it turned out to be another pademelon. The thylacine can again rest in peace.

> No, I would not give you false hope,
> On this strange and mournful day
>
> – Paul Simon, 'Mother and Child Reunion' (1972)

De-extinction of the thylacine

Andrew Pask

De-extinction is the science of bringing an extinct animal back to life, using advanced genetic and reproductive technologies. The concept of de-extinction was science fiction until recently when key advances in our ability to sequence genomes and perform large-scale DNA editing have made this a very real possibility. The question is no longer can this be done, but when will it be done? And, what will the resulting animals look like?

The first hurdle for any de-extinction project is sequencing the animal's genome, which contains the complete blueprint of how to build your extinct species. This was achieved for the thylacine in 2019 and represents one of the best (most complete) extinct genomes sequenced. However, there is a gap, and one which science is yet to bridge, between defining the genome of an extinct animal and bringing that animal back to life.

The first challenge in de-extinction is the concept of what constitutes a sequenced genome. No-one has achieved the complete genome sequence for any vertebrate species. Large holes in the genome still exist in almost all genome assemblies and how these regions contribute to the overall identity of the species is unknown. These gaps, or holes, in genome assemblies lie largely within massive stretches of repetitive DNA. When we sequence a genome we generate lots of small pieces, like the pieces of a puzzle. Some parts of the genome look unique, like distinct parts of a puzzle picture, and can be easily matched to their corresponding puzzle piece, but in the mammalian genome, at least half of the puzzle picture is just repeats. The pieces look very similar (like blue sky) and it is very difficult, and in many instances impossible, to determine what goes where. These repetitive parts of the genome include the centromeres, the narrow constriction at the centre of the chromosomes that controls how the chromosomes segregate during cell division. They also include the telomeres, which are the ends of the chromosomes and perhaps most importantly, the sex chromosomes. The Y-chromosome in particular is incredibly rich in repeats and is absolutely essential in mammals for triggering male development and sperm production. Obviously, in any de-extinction project these regions of the genome are absolutely essential for constructing a viable animal that can reproduce.

The only feasible option to contend with the problem of missing genome sequences is to use an existing functional genome from an extant (living) species as a template and create a hybrid-like genome through genome engineering. From sequencing the thylacine genome we have determined that its closest living relative is the dunnart. To recreate a thylacine you would need to compare its genome to that of the dunnart and identify all the differences. Then, starting with living dunnart cells in culture, you would edit their genome to look like thylacine DNA and use the dunnart repeat sequences to fill in any gaps. So, is this really resurrecting an exact replica of the extinct species? The answer is no, but it may be able to produce an animal that is very close in appearance and behaviour to the extinct one.

The next challenge is how to turn your engineered thylacine-like cells into a living organism. This uses the principals of cloning and assisted reproductive technologies (ART: methods developed to form an embryo and transfer it to a recipient host female and have her carry that pregnancy). This is no mean feat even in commonly used laboratory species such mice and is especially challenging in other species. These technologies are still sorely lacking for marsupials. ART have enormous potential in marsupials to help conservation efforts and so the development of these techniques is ongoing. If they are perfected, these could then be applied to turning an engineered thylacine cell into a viable individual.

Once the 'thylacine' is born you have a single individual in your population. Genetic variation, or the difference in genome sequence between individuals in a given population, is essential for health and survival. It is this genetic variation that enables us to evade disease. Genetic variation is not well understood at the individual level but would need to be engineered into a cloned animal population to ensure its ability to not only survive but thrive back in the ecosystem. A great example of the inherent risk of limited genetic diversity is the Tasmanian devil facial tumour virus that devastated a huge percentage of the devil population because of the lack of genetic variation in this species. Another concern is whether we still have existing suitable habitats for reintroduction of these species. In the case of the thylacine, such viable environments do still exist. However, the current amount of effort required to produce a population of animals large enough to try a reintroduction strategy is untenable, yet as DNA editing technologies continue to improve and the cost of such endeavours reduces, we must be prepared to have conversations in the future about whether we should attempt to de-extinct this unique and enigmatic animal.

What the future holds for the thylacine

Col Bailey

The designation 'extinct' is a truly sombre attestation, denoting the permanent loss of a species. Once sentenced, seldom does a species recover its extant status. The thylacine is one such species. Many today would assert that even a momentary glimpse of a living, breathing thylacine was the equivalent of winning the lottery. A fair assumption perhaps, but let us not forget that some people do actually win.

I am well known for my enduring belief in the thylacine's continued survival. My interest in the thylacine now extends over 50 years and along the way I have spoken with hundreds claiming to have had encounters, of one form or another, with this mysterious marsupial carnivore and they can't all be wrong – or can they?

Science obstinately tells us the thylacine is extinct and has been since the night of 7 September 1936 when the world's last known captive specimen died in Hobart's Domain Zoo. Of course, no-one really knows or can be sure that this animal was the world's last thylacine. I know for certain that it was not.

Early on a cool autumn morning in March 1995, I came face to face with a living, breathing thylacine along the Snake River in the Weld Valley of southern Tasmania. It was as much a shock to me as I was to it and after carefully observing the animal for ~45 seconds I was in no doubt as to what it was. This was the animal I had been researching for the past 28 years and there was no way I could have been mistaken. For the next 17 years I told not a living soul of my discovery, despite being prodded and probed by so many individuals, academics and television documentary crews to do so.

In continually declaring the species extant my views became widely known, but they were not generally accepted by the scientific community, in which there were many with strong opinions on a subject they so often knew little or anything about.

As an enthusiastic young researcher, I sought out some of the old bushmen who had actively both tracked and trapped the thylacine. Their working knowledge of the animal was second to none and it was from these men that I learnt so much. These were the true experts on the thylacine and there are none alive today. To be a true expert one must know every single thing there is to know about their subject – and there is so much we do not know about the thylacine. My mentor and dear friend, the late Professor Eric Guiler (1922–2008), one of the world's acknowledged authorities on the thylacine, often so wisely proclaimed: 'We just don't know – there is so much we still don't know about the thylacine'.

So, what does the future hold for the thylacine? First, we have to prove the species still exists and short of a live or freshly dead specimen, a photograph authenticated by known authorities in the species may be our only hope. Numerous search parties have, over the years, spent much time and money trying to bring this elusive beast to ground without success. There is, even now, a highly geared up and extremely knowledgeable unit at work in the field and no doubt a dozen others pitting their limited knowledge and experience against an animal that surely must know every trick in the book. However, I am hopeful that one of these groups will be successful in obtaining the necessary proof that the thylacine is still extant.

If the thylacine has a future in Tasmania, and I sincerely believe that it has, we have no time to lose. The species is critically endangered, with probably fewer than 100 individuals left roaming the remote wilderness of Tasmania. The sands of the hour glass are running out and we have little time left to prove the species still exists before its memory is consigned to the past.

PART 9: BEYOND REALITY

Narrating perseverance: an overview of thylacines in fiction

Daisy Ahlstone

That's when I will be truly dead – when I exist in no one's memory.

– Irvin D Yalom (1989)

Writing is an opportunity for reimagining our world. In writing the story of the thylacine, authors not only keep its memory alive, but give it new life through narration. Any writing of the thylacine sustains its symbolic power in a tangible, material form.

These narrations come in a variety of genres: scientific and academic texts that theorise the thylacine's ecological niche; speculative fiction that imagines its presence and absence; non-academic educational works, such as children's literature, which teach the importance of understanding extinction; oral histories that recall the thylacine's presence; film that reimagines its image; fanfiction and other unpublished works that blur the boundaries of language and art. Sometimes this work is formally published and other times it is simply posted online to share with other thylacine fans. The thylacine's intriguing place in history, charismatic body and imposed personality, the dangerous legends and those of its survival today, have created a sort of mythos, a fan-community, for the thylacine.

Scientific and academic literature has worked to debunk much of the historically inaccurate information and uncharitable and demonising depictions of thylacines that influenced their extinction. In her books, Carol Freeman (2010, 2014) draws a linear history in the depiction of the thylacine through scientific illustration and institutional representations, analysing the artwork's cultural impact and significance. Hers and other works, while working in formalised institutional contexts, help to write a history of the thylacine that does justice to its cultural and ecological significance to the world.

Thylacine fiction writing has directly assisted in generating global interest. Authors and readers of fiction describing thylacine search parties in the Tasmanian wilderness are not themselves experiencing the pursuit, but they are perhaps embodying the feelings and excitement of their characters making this monumental discovery. Popular educational literature featuring the thylacine is also widely read. The main goal of this genre is often to warn and teach people across generations about extinction and the story of the thylacine from a biological perspective. Occasionally, these texts include a direct discussion, or at least an implication of human influence on the thylacine's extinction.

Another common type of thylacine writing is collections of oral histories: personal narratives written down describing first-person encounters with thylacines. Particularly prolific in this genre is author and researcher Col Bailey, who has dedicated his life to documenting oral histories of thylacine encounters from the individuals themselves (Bailey 2001, 2013, 2016; see pp. 86–8). These oral histories give us hope for the possibility of seeing a thylacine in the wild today and allow us to imagine ourselves as directly connected to its ecological legacy.

Other writings about thylacines existing online are a more imaginative blend of non-fiction, art and speculative fiction. This folk narrative writing (informally published thylacine fanfiction) is most exciting to me as a folklorist. These works seek to educate the public about this special animal, human's role in its extinction and what humans can now do to help. A unique example of this creation is Morgan Mudway's *Stripes Ascending* (Mudway 2019), which includes a graphic retelling of the thylacine's tale alongside interactive components in the text (word games, puzzles, and colouring pages). This fanfiction is typically not monetised and is generated out of personal interest in the thylacine, although the texts are occasionally supported through online crowd-funding. Not unlike this book, many of these community generated texts have a charitable component tied to Tasmanian devil conservation. The thylacine fandom not only cares about the thylacine, but also the protection of all Australian native species threatened with extinction.

As a more visually stimulating narrative, the thylacine has made a marginal impact in cinema. At this point in your reading, you have probably been exposed to the infamous book and later film *The Hunter*, both of which received international success. The only other fictional film of note is titled *The Howling III: The Marsupials* (1987), the third horror film in an eight-film series documenting werewolves across the world. This particular film focuses on *werelacines* or marsupial–human crossbreeds found only in Australia. Television programs that play with these more negative representations of thylacines include *Monster Quest*, *Extinct or Alive* and *Expedition Unknown*. Crowd-funded, locally produced documentaries work against some of the more harmful thylacine stereotypes and present the most current research on the possibility of their existence today.

Regardless of whether they are depicted in formal or informal spaces, the simple act of depicting the thylacine at all gives it continued life; the legacy of its present existence lives on in these material forms.

Gaming extinction: representation of the thylacine in video games

Daisy Ahlstone

Video games open unique doorways into the representation and history of the thylacine as an extinct species. The continued rise of video games' cultural importance is found in 38% of the global population and accounts for more than 159 billion US dollars, making it one of the most lucrative entertainment markets (Clement 2021). Representations of thylacines in those video games, as expected, are rare but notable, dating back at least to the early 2000s. The video game industry combines artwork and storytelling, while allowing the player a direct role in influencing the story. This makes thylacine representation in video games unique among representations of the species.

Several factors converged in the late 1990s that help explain the presence of a recently extinct animal in so many games. The popularity of video games such as 'Super Mario Bro's', 'Pokémon', 'Zelda' and the 'Final Fantasy' series attached a generation to storytelling through an active and participatory role in the unfolding of narratives. Additionally, the science fiction boom of 1990s films, such as 'Jurassic Park', primed the public imagination to engage with emerging technologies related to the extension or resurgence of life such as through cloning. The research conducted by Professor Mike Archer in the late 1990s infused the thylacine's image into the public imagination through this lens of fantastical science. The link between storytelling, popular science and video games became a ripe environment for the public's science fiction fantasies to expropriate the thylacine's story and image. All of this combined with its representation in other pop culture genres gave the thylacine a lasting image in culture, which is explored through many mediums, particularly video games.

Before jumping into the thylacine's representation, it is important to clarify video game terminology, as thylacines are typically found in less standard video game forms, with a few notable exceptions. The actual gameplay can come in many forms, but for thylacines, the most notable representation is the plot-driven platformer game 'TY the Tasmanian Tiger'. A platformer game is a subgenre of action-based video games, which utilise running, jumping and climbing as the main method of gameplay. This subgenre of game is most famously represented by Nintendo's 'Mario' franchise.

Most other thylacine representations in video games are found in less common forms, such as downloadable content (DLC), modifications or 'mods' and indie games. DLCs are produced by video game companies and typically require additional purchases or downloads beyond the main game; they are released to keep people interested in the games and building more exciting content while profiting long-term. Sometimes fan-created mods are purchased or supported by institutions and turned into official DLCs. An example of the thylacine packaged in this way is the 'Predator DLC' released by Ubisoft for 'Far Cry 3', which introduced several endangered or extinct creatures into the main game for the player to hunt.

Additionally, skilled and dedicated fans can write or edit a video game's code to modify the way it is looked or played. These mods are created and uploaded to websites for other players to download. Well-designed mods are highly sought after; in the case of those that include thylacines, the players are seeking to add extinct animals, charismatic animals or a thylacine to the game. For example, the incredibly popular game 'Roblox' has a downloadable mod called, 'Cenozoic survival' that features a thylacine as an additional character to interact with.

Thylacines are depicted in video games in two main ways, with a notable exception: those to be hunted or killed, those to be protected and in the important exception, as the main character ('TY the Tasmanian Tiger'). In the first two scenarios, the assumption is that the player must find the living thylacine and decide to kill, protect or hide their knowledge about the species. Occasionally, thylacines are featured as side or tertiary characters (such as Tiny Tiger, a mutated villain in the 'Crash Bandicoot' series). In most other games, the thylacine is not featured as a central character, but is rather introduced into the game as a downloadable mod. Representation of the thylacine in video games is most often about negotiating presence and plays into the legend of its present existence in Tasmania today.

By far the most famous thylacine-related video game is 'TY the Tasmanian Tiger', published in 2002 by Krome Studios. The game features Ty, a thylacine, as the main playable character. He uses a boomerang as a weapon on his mission to find 'Thundereggs' to save his family trapped in another dimension called 'the Dreaming'. 'TY the Tasmanian Tiger' sold over a million copies worldwide in the year of its release, likely due in part to its similarities with the even more popular 'Mario' series. The game has multiple instalments and updated versions and even a new edition for the Nintendo Switch released in 2020. It is likely one of the first places many young non-Australians are exposed to thylacines.

Another example of a videogame prominently featuring the thylacine is the independent game 'Petra and the Thylacine' launched in 2020. In this point-and-click style game, the player literally points and clicks the cursor on way to the next layer of the plot from several options to follow along a narrative that can change depending on the player's choices. This game follows the fairy-tale formula of 'Little Red Riding Hood': players click their way along a visual adventure in the Cataract Gorge of Tasmania on the way to find grandma's house, only to have some surprises along the way. The thylacine is not revealed to be a main character at first, but rather, in a blend between legend and fairy-tale, the player's game-companion, Petra, transforms into a thylacine depending on which ending the story takes. The game also incorporates stylised photography from Tasmania and historic, geographic and cultural information about the island state. This demonstrates the significance the thylacine has in understanding the role of place and space at play in this retelling of 'Little Red Riding Hood'.

A popular example of a thylacine included in a DLC is the 'Predator DLC' released for 'Far Cry 3' in 2012, which follows a young American as he fights against modern-day pirates in a fictionalised Pacific island. One of the significant side activities includes hunting endangered species from the South Pacific for skins to trade for better guns and ammunition. The thylacine's purpose in this DLC is exclusively to be killed by the main character for sport. The 'Far Cry' Fandom Wiki page describes the significance of the legend of the thylacine's present existence as a main factor for their inclusion in the game. In fact, the game's internal 'survival guide', which serves as a handbook for the player, includes this entry written from the perspective of a side character, Willis Huntley, who states:

> Thylacines are also known as Tasmanian tigers. Zoologists say that these creatures went extinct due to aggressive hunting, while I bet that the zoologists made them up in the first place. As for the tigers on the island? They aren't real (Thylacine 2020a).

In the actual gameplay, the thylacine has low health, is timid and hesitates to attack the player, which makes them easy to kill. The 'Far Cry' series itself represents a 'first-person shooter' style game, where killing and fighting is the primary mode of interaction in the virtual world. Its presentation of the thylacine is no exception, as killing thylacines is an additional way to make money and gain resources. Thylacines often run away from the main character when spotted, and travel in small groups of three in this DLC. Their inclusion reflects the pop culture story of the thylacine: an animal hunted to extinction long ago that still lives an unconfirmed existence in the wild today.

The thylacine is also seen in the 'ZooTycoon' series, making an appearance in ZooTycoon 2: Extinct Animals (Thylacine 2020b). The legend of its existence beyond extinction again plays a huge part in their description and justification for including the iconic creature in this series. In 'ZooTycoon 2', players have created an informational wiki site to assist other players with the game; it is entirely created and administered by fans on a voluntary basis. The ZooTycoon fandom site claims the thylacine became extinct due to overhunting, but that their decline was also affected by competition with dingoes and that people still readily report seeing thylacine-like animals in the wild today. The goal of 'ZooTycoon 2' is to run a successful zoo, both for animals and patrons. The thylacine's role is simply as a species the player may create a successful enclosure for, along with other animals, extinct and extant, to build a successful zoo. This expansion pack added other extinct animals as well, including woolly rhinos, mastodons, a range of dinosaurs and the dodo. Its association with formerly extinct animals is interesting, considering the legend of the thylacine's present existence is noted as one of the most intriguing parts of the animal, justifying its inclusion in the game.

Video game mods are a key part of understanding the ways fans interact with video games, outside of gameplay, in fan-communities. If they feel something is absent in the game, they can write code to include it in the game themselves. This is relatively straightforward in the case of modding a thylacine into a game, because it typically involves putting a new colouration on an existing dog mechanic to make it look similar to a thylacine or adding a single new character without dialogue or much complicated player interaction within the game. An example of a thylacine mod in a video game is the texture mods for 'Roblox' and 'Minecraft', which allow the player to make dogs or wolves that would otherwise appear in the video games look like thylacines. The characters function the same way they would in the game normally, with the added fun of appearing like thylacines instead of their original, official game appearance.

Mods are especially significant because their vernacular nature requires a player to seek out or create the content, making their inclusion and creation very intentional and fan-generated. The thylacine's cult popularity and application to existing game designs make it possible for fans to create simple mods of thylacines for the games they already love to celebrate their interest in the animal and add more whimsy to games they already love. In a way, playing a video game with a thylacine included allows the player to represent what it could be like to have thylacines existing still today, playing into their most popular legend.

The thylacine is a niche animal and a species that global audiences are not well acquainted with, which makes the existence of these mods especially intriguing, because they imply a person interested in the thylacine cares enough to actively seek out and create modifications to main storylines. The thylacine continues to capture the imagination of people across mediums, including video games. The dynamic of the thylacine's legend of living past its official extinction is replicated in the search for the thylacine through video games.

Both as I write this, and as other players seek out the thylacine in video games, the animal maintains allusivity and will only be found by those who seek out its existence through personal interest.

The thylacine is represented in just about every form of popular culture. The important thing to recognise about the thylacine's representation in video games is the tropes pulled from both its historical past and the legends of present existence. These narratives have seeped into common knowledge partly because of pop culture representations, both in Australia and beyond, and play into the liminality of the thylacine in stories today.

A day in the life of a thylacine keeper

Nicole Dyble

Date: 23 December 2036.
Weather conditions: warm, wind increasing in late afternoon.

Jobs completed today

1. Scat collection, including internal game trails and latrines.
2. Internal and external perimeter check of main enclosure and Isolation pens 1 and 2.
3. Ponds in main enclosure scrubbed, drained and refilled.
4. Water bowls cleaned and filled in Iso pens.
5. Collected and swapped out SD cards in cameras for review (need to recharge batteries).
6. Hosed off 'furniture' (granite blocks) and climbing structures.
7. Covered exposed wire along southern fence line in main enclosure with gravel.
8. Took extra browse into Iso 2 for boys, covered the artificial denning boxes and added some natural den sites.
9. Sorted through the 'dry store' in pantry, as well as checking medical supplies and enrichments.
10. Cleaned meat prep station with high-pressure hose, cleaned whitegoods in meat prep area.

Animals sighted

- Coorinna and joeys sun basking in Iso pen 1 during service. When keepers entered, joeys retreated and hid. Coorinna retreated but watched. Head keeper kept an eye on her while second keeper did scat collection and cleaned water sources. Although wary, she didn't approach.
- 2 Males in Iso 2 weren't reactive to keeper presence. Although they do have different behaviour profiles, the larger male a little more confident and curious generally. Especially during cleaning routines.
- The female and one of the males in the main enclosure came out to sniff and watch keepers do maintenance on the exposed wire during the day, but no sign of the

second male. (If not seen during feeding, will check cameras for movement tomorrow morning.)

Feeding/enrichments

Feeding approximately 1.5 kg of wallaby per thylacine tonight. Some bone and fur included. Smaller pieces also scattered for Coorinna and her young. Tomorrow will be a fast day except for Coorinna and joeys, who are to be given a high-fat mince mix with eggs, grated cheese, carnivore powder and vegetable oil. Please give enrichments tomorrow of nutrigel in egg carton pods, scattered and placed up into climbing structures and on rock furniture. Also scats from neighbouring pens to be scattered for enrichment. Possum to be fed the following day (higher in fat).

Notes for tomorrow

Please observe if any food left uneaten. Look for worm eggs in scats and check SD cards. Body condition scores are due to be recorded (Plate 19).

Shopping list

1. Carnivore powder
2. Kelp powder
3. Eggs
4. Beef stock powder
5. Drontal worming tabs

References

Abbie AA (1940) The excitable cortex in *Perameles, Sarcophilus, Dasyurus, Trichosurus* and *Wallabia* (*Macropus*). *Journal of Comparative Neurology* **72**, 469–487. doi:10.1002/cne.900720303

Abbott E (1864) The English and Australian Cookery Book: Cookery for the Many, as well as the "Upper Ten Thousand". Sampson Low, Son, and Marston, London.

Advocate (1923) Broken Hill footballers. Burnie, 19 July.

Advocate (1927) H. S. Aylett and Sons. Burnie, 1 April.

Advocate (1928) A Tasmanian Tiger. Burnie, 20 March.

Advocate (1930a) A Rare Catch. Burnie, 10 July.

Advocate (1930b) Waratah. Rare Catch. Burnie, 11 August.

Advocate (1930c) Our Tiger. Burnie, 22 January.

Advocate (1935) Costly food for zoo animals. Burnie, 23 July.

Advocate (1945) Tasmanian tiger wanted. Burnie, 22 June.

Advocate (1946) Tasmanian Tiger Expedition Fails. Burnie, 22 March.

Allport M (1868) Remarks on Mr. Krefft's 'Notes on the Fauna of Tasmania'. *Papers and Proceedings of the Royal Society of Tasmania* 1868, 33–36.

Andrews P (1990) *Owls, Caves and Fossils*. Natural History Museum Publications, London.

Anonymous (1932) Hand written copy of the annual report of the Curtis Museum dated 3rd December 1932. Curtis Museum Archives, Hampshire, UK.

Archer M (1974) New information about the Quaternary distribution of the thylacine (Marsupialia, Thylacinidae) in Australia. *Journal of the Royal Society of Western Australia* **57**, 43–50.

Archer M (1982) A review of Miocene thylacinids (Thylacinidae, Marsupialia), the phylogenetic position of the Thylacinidae and the problem of apriorisms in character analysis. In *Carnivorous Marsupials*. (Ed. M Archer) pp. 445–476. Royal Zoological Society of New South Wales, Sydney.

Archer M, Hand SJ, Black KH, Beck RMD, Arena DA, *et al.* (2016) A new family of bizarre durophagous carnivorous marsupials from Miocene deposits in the Riversleigh World Heritage Area, northwestern Queensland. *Scientific Reports* **6**, 26911. doi:10.1038/srep26911

Attard MRG (2013) Who's on the menu: marsupial carnivore feeding ecology and extinction risk. PhD thesis. University of New South Wales, Australia.

Attard MRG, Chamoli U, Ferrara TL, Rogers TL, Wroe S (2011) Skull mechanics and implications for feeding behaviour in a large marsupial carnivore guild: the thylacine, Tasmanian devil and spotted-tailed quoll. *Journal of Zoology* **285**, 292–300. doi:10.1111/j.1469-7998.2011.00844.x

Attard MRG, Parr WCH, Wilson LAB, Archer M, Hand SJ, *et al.* (2014) Virtual reconstruction and prey size preference in the mid Cenozoic thylacinid, *Nimbacinus dicksoni* (Thylacinidae, Marsupialia). *PLoS ONE* **9**, e93088. doi:10.1371/journal.pone.0093088

Australasian Post (1956) Extinct? Melbourne, 16 August.

Badenhorst S, Plug I (2011) Unidentified specimens in zooarchaeology. *Palaeontologia Africana* **46**, 89–92.

Bailey C (2001) *Tiger Tales: Stories of the Tasmanian Tiger*. HarperCollins, Sydney.

Bailey C (2013) *Shadow of the Thylacine: One Man's Epic Search for the Tasmanian Tiger*. Five Mile Press, Scoresby.

Bailey C (2016) *Lure of the Thylacine: True Stories and Legendary Tales of the Tasmanian Tiger*. Echo Publishing, Melbourne.

Banks PB, Hochuli DF (2017) Extinction, de-extinction and conservation: a dangerous mix of ideas. *Australian Zoologist* **38**, 390–394. doi:10.7882/AZ.2016.012

Barnes MA, Turner CR, Jerde CL, Renshaw MA, Chadderton WL, *et al.* (2014) Environmental conditions influence eDNA persistence in aquatic systems. *Environmental Science & Technology* **48**, 1819–1827. doi:10.1021/es404734p

Barrett C (1944) *Isle of Mountains*. Cassell, Melbourne.

Barrington G (1810) *The History of New South Wales* 2nd edn. Sherwood, Neely and Jones, London.

Baynes A, Jones B (1993) The mammals of Cape Range peninsula, north-western Australia. *Records of the Western Australian Museum* (Supplement 45), 207–225.

Beck RMD (2013) A peculiar faunivorous metatherian from the early Eocene of Australia. *Acta Palaeontologica Polonica* **60**, 123–129. doi:10.4202/app.2013.0011

Belov K, Sanderson CE, Deakin JE, Wong ESW, Assange D, *et al.* (2007) Characterization of the opossum immune genome provides insight into the evolution of the mammalian immune system. *Genome Research* **17**, 982–991. doi:10.1101/gr.6121807

Berns GS, Ashwell KWS (2017) Reconstruction of the cortical maps of the Tasmanian tiger and comparison to the Tasmanian devil. *PLoS ONE* **12**, e0168993. doi:10.1371/journal.pone.0168993

Berry O, Sarre SD, Farrington L, Aitken N (2007) Faecal DNA detection of invasive species: the case of feral foxes in Tasmania. *Wildlife Research* **34**, 1–7. doi:10.1071/WR06082

Beveridge I, Spratt DM (2015) Biodiversity and parasites of wildlife: helminths of Australasian marsupials. *Trends in Parasitology* **31**, 142–148. doi:10.1016/j.pt.2014.10.007

Blackwell W (unpublished) 'Thylacine report NS1143/1/17'. Queen Victoria Museum and Art Gallery, Launceston, 27 November 1951.

Bonaparte CL (1832) *Saggio d'una Distribuzione Metodica degli Animali Vertebrati a Sangue Freddo.* Presso Antonio Boulzaler, Rome.

Borthwick CR, Young LJ, Old JM (2014) Development of the immune tissues and in marsupial pouch young. *Journal of Morphology* **275**, 822–839. doi:10.1002/jmor.20250

Brain WA (1928a) City of Hobart. *The Mercury*, Hobart, 13 June.

Brain WA (1928b) City of Hobart. *Examiner*, Launceston, 16 June.

Brandl EJ (1972) Thylacine designs in Arnhem Land rock paintings. *Archaeology & Physical Anthropology in Oceania* **7**, 24–30.

Brook BW, Sleightholme SR, Campbell CR, Jarić I, Buettel JC (2021) Resolving when (and where) the Thylacine went extinct. *bioRxiv* doi:10.1101/2021.01.18.427214

Brown OFF (2006) Tasmanian devil (*Sarcophilus harrisii*) extinction on the Australian mainland in the mid-Holocene: multicausality and ENSO intensification. *Alcheringa: An Australasian Journal of Palaeontology* **30**, 49–57. doi:10.1080/03115510609506855

Brüniche-Olsen, A, Jones, ME, Burridge, CP, Murchison, EP, Holland, BR, *et al.* (2018) Ancient DNA tracks the mainland extinction and island survival of the Tasmanian devil. *Journal of Biogeography* **45**, 963–976. doi:10.1111/jbi.13214

Bryant SL (1988) Seasonal breeding in the eastern quoll, *Dasyurus viverrinus* (Marsupialia: Dasyuridae). PhD thesis. The University of Tasmania, Australia.

Bullock WA (1808) *A Companion to the Liverpool Museum*, 6th edn. Ferraby, Liverpool.

Bullock WA (1810a) *A Companion to Mr. Bullock's Museum,* 8th edn. H. Reynell & Son, London.

Bullock WA (1810b) *A Companion to Mr. Bullock's Museum,* 9th edn. H. Reynell, London.

Bullock WA (1811) *A Companion to Mr. Bullock's Museum,* 10th edn. H. Reynell, London.

Bullock WA (1812) *A Companion to Mr. Bullock's London Museum and Pantherion.* Reynell, London.

Bulmer S (1966) The prehistory of the New Guinea Highlands: a discussion of archaeological field survey and excavations, 1959–60. MA thesis. University of Auckland, Auckland.

Burbidge AA, McKenzie NL (1989) Patterns in the modern decline of western Australia's vertebrate fauna: causes and conservation implications. *Biological Conservation* 50, 143–198. doi:10.1016/0006-3207(89)90009-8

Burbidge AA, Johnson K, Fuller P, Southgate R (1988) Aboriginal knowledge of the mammals of the central deserts of Australia. *Australian Wildlife Research* **15,** 9–39. doi:10.1071/WR9880009

Burnett GT (1830) Illustrations of the Quadrupeda, or Quadrupeds, being the arrangement of the true four-footed Beasts indicated in outline. *Quarterly Journal of Science, Literature, and Art* **28**, 336–353.

Camens AB, Carey SP, Arnold LJ (2017) Vertebrate trace fossils from the Late Pleistocene of Kangaroo Island, South Australia. *Ichnos* **25**, 232–251. doi:10.1080/10420940.2017.1337633

Campbell CR (1999) TheThylacine in Captivity:Zoos, Menageries and Circuses – Tasmania. The Thylacine Museum, Fort Worth. <http://www.naturalworlds.org/thylacine/captivity/zoos/Tasmania/Tasmania_3.htm>.

Campbell CR (2021) History: Extinction versus Survival. The Thylacine Museum, Fort Worth. <http://www.naturalworlds.org/thylacine/history/extvssurv/extinction_vs_survival_14.html>.

Carlson CJ, Hopkins SR, Bell KC, Doña J, Godfrey SS, *et al.* (2020) A global parasite conservation plan. *Biological Conservation* **250**, 108596. doi:10.1016/j.biocon.2020.108596

Carroll SB (2008) Evo-devo and an expanding evolutionary synthesis: a genetic theory of morphological evolution. *Cell* **134**, 25–36. doi:10.1016/j.cell.2008.06.030

Cartmill M, Brown K, Atkinson C, Cartmill EA, Findley E, *et al.* (2020) The gaits of marsupials and the evolution of diagonal-sequence walking in primates. *American Journal of Physical Anthropology* **171**, 182–197. doi:10.1002/ajpa.23959

Cartmill M, Lemelin P, Schmitt D (2002) Support polygons and symmetrical gaits in mammals. *Zoological Journal of the Linnean Society* **136**, 401–420. doi:10.1046/j.1096-3642.2002.00038.x

Case JA (1985) Differences in prey utilization by Pleistocene marsupial carnivores, *Thylacoleo carnifex* (Thylacoleonidae) and *Thylacinus cynocephalus* (Thylacinidae). *Australian Mammalogy* **9**, 45–52.

Castro-Arellano I, Madrid-Luna C, Lacher TE, León-Paniagua L (2008) Hair-trap efficacy for detecting mammalian carnivores in the tropics. *Journal of Wildlife Management* **72**, 1405–1412. doi:10.2193/2007-476

Chaloupka G (1993) *Journey in Time*. Reed Books, Sydney.

Chaloupka G, Murray P (1986) Dreamtime or reality? Reply to Lewis. *Archaeology in Oceania* **21**, 145–147. doi:10.1002/j.1834-4453.1986.tb00136.x

Chaplin A (1954) Unpublished letter. Cited in Paddle R (2000) *The Last Tasmanian Tiger: The History and Extinction of the Thylacine*. p. 191. Cambridge University Press, Cambridge.

Chisholm AH (1924) Ways of the wild. *The Daily Telegraph*, Sydney, 19 December.

Chisholm AH (1936) Bush notes. *The Australasian*, Melbourne, 3 October.

Chisholm AH (1945) In quest of the Tasmanian "tiger". *The Herald*, Melbourne, 30 October.

Chronicle (1949) Is tiger on the prowl again? Adelaide, 21 July.

Cifelli RL, Davis BM (2015) Tribosphenic mammals from the Lower Cretaceous Cloverly Formation of Montana and Wyoming. *Journal of Vertebrate Paleontology* **35**, e920848. doi:10.1080/02724634.2014.920848

Circular Head Chronicle (1926a) Hyena choked in snare at Salmon River. Smithton, 11 August.

Circular Head Chronicle (1926b) Hand bitten by a hyena. Smithton, 29 September.

Circular Head Chronicle (1927) Montagu. Stanley, 23 November.

Circular Head Chronicle (1929) Arthur River. Stanley, 29 May.

Clarkson C, Jacobs Z, Marwick B, Fullagar R, Wallis L, *et al.* (2017) Human occupation of northern Australia by 65,000 years ago. *Nature* **547**, 306–310. doi:10.1038/nature22968

Clarkson C, Smith M, Marwick B, Fullagar R, Wallis LA, *et al.* (2015) The archaeology, chronology and stratigraphy of Madjedbebe (Malakunanja II): a site in northern Australia with early occupation. *Journal of Human Evolution* **83**, 46–64. doi:10.1016/j.jhevol.2015.03.014

Clement J (2021) Number of gamers worldwide 2023, Statista. <www.statista.com/statistics/748044/number-video-gamers-world/>.

Colbron-Pearse D (1937) Letters to the editor. *The Mercury*, Hobart, 4 January.

Colonial Times (1833) Hobart, 25 June.

Colonial Times (1834) Hobart, 9 September.

Colonial Times (1841) Hobart, 14 September.

Coplestone R (1926) NE early days: reminiscences. *North Eastern Advertiser*, Scottsdale, 10 December.

Corbett LK (1995) *The Dingo in Australia and Asia*. Cornell University Press, New York.

Cornwall Chronicle (1879) Kangaroo hunting. Launceston, 31 January.

Cowart DA, Murphy KR, Cheng CHC (2018) Metagenomic sequencing of environmental DNA reveals marine faunal assemblages from the West Antarctic Peninsula. *Marine Genomics* **37**, 148–160. doi:10.1016/j.margen.2017.11.003

Crisp E (1855) On some points relating to the anatomy of the Tasmanian wolf (*Thylacinus*) and of the Cape hunting dog (*Lycaon pictus*). *Proceedings of the Zoological Society of London* **23**, 188–192.

Cunningham CX, Comte S, McCallum H, Hamilton DG, Hamede R, *et al.* (2021) Quantifying 25 years of disease-caused declines in Tasmanian devil populations: host density drives spatial pathogen spread. *Ecology Letters* **24**, 958–969. doi:10.1111/ele.13703

Cunningham CX, Johnson CN, Barmuta LA, Hollings T, Woehler E, *et al.* (2018) Top carnivore decline has cascading effects on scavengers and carrion persistence. *Proceedings of the Royal Society B: Biological Sciences* **285**, 20181582. doi:10.1098/rspb.2018.1582

Cunningham CX, Johnson CN, Hollings T, Kreger K, Jones ME (2019a) Trophic rewilding establishes a landscape of fear: Tasmanian devil introduction increases risk-sensitive foraging in a key prey species. *Ecography* **42**, 2053–2059. doi:10.1111/ecog.04635

Cunningham CX, Scoleri V, Johnson CN, Jones ME (2019b) Temporal partitioning of activity: rising and falling top-predator abundance triggers community-wide shifts in diel activity. *Ecography* **42**, 2157–2168. doi:10.1111/ecog.04485

Cunningham CX, Johnson CN, Jones ME (2020) A native apex predator limits an invasive mesopredator and protects native prey: Tasmanian devils protecting bandicoots from cats. *Ecology Letters* **23**, 711–721. doi:10.1111/ele.13473

Dabney J, Meyer M, Pääbo S (2013) Ancient DNA damage. *Cold Spring Harbor Perspectives in Biology* **5**, a012567. doi:10.1101/cshperspect.a012567

Daily Mercury (1926) Tasmanian wolf. Mackay, 14 April.

Daniels GD, Kirkpatrick JB (2012) The influence of landscape context on the distribution of flightless mammals in exurban developments. *Landscape and Urban Planning* **104**, 114–123. doi:10.1016/j.landurbplan.2011.10.003

Davies JL (1965) *Atlas of Tasmania.* Lands and Surveys Department, Hobart.

de Labillardière JJH (1800) *Voyage in Search of La Pérouse.* J. Stockdale, London.

Delphin D (1930) Tasmanian tiger. *Advocate*, Burnie, 20, 22, 23, 26, 27 September.

Denham T, Mountain M-J (2016) Resolving some chronological problems at Nombe rock shelter in the highlands of Papua New Guinea. *Archaeology in Oceania* **51**, 73–83. doi:10.1002/arco.5114

Dietzel MA (2016) Sentiment-based predictions of housing market turning points with Google trends. *International Journal of Housing Markets and Analysis* **9**, 108–136. doi:10.1108/IJHMA-12-2014-0058

Dixon J (1989) Thylacinidae. In *Fauna of Australia. Vol. 1B. Mammalia.* (Eds D Walton and B Richardson) pp. 549–559. Australian Government Publishing Service, Canberra.

Doherty K (1977) When we caught a tiger. In *N. W. Tasmania Short Stories and Articles.* (Ed. P Buckby). Hamlet Publishing, Burnie.

Donders TH, Haberle SG, Hope G, Wagner F, Visscher H (2007) Pollen evidence for the transition of the Eastern Australian climate system from the post-glacial to the present-day ENSO mode. *Quaternary Science Reviews* **26**, 1621–1637. doi:10.1016/j.quascirev.2006.11.018

Donders TH, Wagner-Cremer F, Visscher H (2008) Integration of proxy data and model scenarios for the mid-Holocene onset of modern ENSO variability. *Quaternary Science Reviews* **27**, 571–579. doi:10.1016/j.quascirev.2007.11.010

Dooley R (2017) From 'abundance of emues' to a rare bird in the land: the extinction of the Tasmanian emu. *Papers and Proceedings: Tasmanian Historical Research Association* **64**, 4–17.

Driessen MM, Hocking GJ (1992) *Review and Analysis of Spotlight Surveys in Tasmania: 1975–1990.* Department of Parks, Wildlife and Heritage, Tasmania.

Duchêne DA, Bragg JG, Duchêne S, Neaves LE, Potter S, *et al.* (2018) Analysis of phylogenomic tree space resolves relationships among marsupial families. *Systematic Biology* **67**, 400–412. doi:10.1093/sysbio/syx076

Dunnet GM, Mardon DK (1974) A monograph of Australian fleas (Siphonaptera). *Australian Journal of Zoology Supplementary Series* **22**, 1–273. doi:10.1071/AJZS030

Edwards J (unpublished) List of thylacines at London Zoo, held by Zoological Society of London, London 1996.

Eldridge MDB, Beck RMD, Croft DA, Travouillon KJ, Fox BJ (2019) An emerging consensus in the evolution, phylogeny, and systematics of marsupials and their fossil relatives (Metatheria). *Journal of Mammalogy* **100**, 802–837. doi:10.1093/jmammal/gyz018

Engelman RK, Anaya F, Croft DA (2017) New palaeothentid marsupials (Paucituberculata) from the middle Miocene of Quebrada Honda, Bolivia, and their implications for the palaeoecology, decline and extinction of the Palaeothentoidea. *Journal of Systematic Palaeontology* **15**, 787–820. doi:10.1080/14772019.2016.1240112

Epstein B, Jones M, Hamede R, Hendricks S, McCallum H, *et al.* (2016) Rapid evolutionary response to a transmissible cancer in Tasmanian devils. *Nature Communications* **7**, 12684. doi:10.1038/ncomms12684

Evans N (2018) *Words of Life.* Little Toller Books, Dorset, UK. <https://www.littletoller.co.uk/the-clearing/poetry/words-of-life-by-nicholas-evans>.

Examiner (1909) Stanley. Launceston, 7 July.

Examiner (1912) A 5ft. tiger. Launceston, 24 May.

Examiner (1920) St Mary's. Launceston, 18 June.

Examiner (1922) Protecting native fauna. Launceston, 18 October.

Fancourt BA, Hawkins CE, Cameron EZ, Jones ME, Nicol SC (2015) Devil declines and catastrophic cascades: is mesopredator release of feral cats inhibiting recovery of the eastern quoll? *PLoS ONE* **10**, e0119303. doi:10.1371/journal.pone.0119303

Feigin CY, Newton AH, Doronina L, Schmitz J, Hipsley CA, *et al.* (2018) Genome of the Tasmanian tiger provides insights into the evolution and demography of an extinct marsupial carnivore. *Nature Ecology & Evolution* **2**, 182–192. doi:10.1038/s41559-017-0417-y

Feigin CY, Newton AH, Pask AJ (2019) Widespread cis-regulatory convergence between the extinct Tasmanian tiger and gray wolf. *Genome Research* **29**, 1648–1658. doi: 10.1101/gr.244251.118

Figueirido B, Janis CM (2011) The predatory behaviour of the thylacine: Tasmanian tiger or marsupial wolf? *Biology Letters* **7**, 937–940. doi:10.1098/rsbl.2011.0364

Fillios M, Crowther MS, Letnic M (2012) The impact of the dingo on the thylacine in Holocene Australia. *World Archaeology* **44**, 118–134. doi:10.1080/00438243.2012.646112

Finlayson HH (1935) *The Red Centre: Man and Beast in the Heart of Australia.* Angus and Robertson, Sydney.

Finnane K (2011) Pmere Arntarntareme/Watching This Place. *Artlink* **31**, 89.

Fischer EM, Knutti R (2014) Heated debate on cold weather. *Nature Climate Change* **4**, 537–538. doi:10.1038/nclimate2286

Fisher DO (2010) Trajectories from extinction: where are missing mammals rediscovered? *Global Ecology and Biogeography* **20**, 415–425. doi:10.1111/j.1466-8238.2010.00624.x

Fisher DO, Blomberg SP (2011) Correlates of rediscovery and the detectability of extinction in mammals. *Proceedings of the Royal Society B: Biological Sciences* **278**, 1090–1097. doi:10.1098/rspb.2010.1579

Fiske P, Vasseleu C (2022) *Tiger on the Rocks*, documentary. Directed by C Vasseleu. Bower Bird Films.

Fitzgerald GL (2011) *Prisoners, Pioneers, and a Prospecting Preacher.* Self-Published, Melbourne.

Flannery TF (1995) *Mammals of New Guinea.* Reed Books, Sydney.

Flannery TF, Mountain M-J, Aplin K (1983) Quaternary kangaroos (Macropodidae: Marsupialia) from Nombe rock shelter, Papua New Guinea: with comments on the nature of the megafaunal extinction in the New Guinea Highlands. *Proceedings of the Linnean Society of New South Wales* **107**, 77–97.

Fleay DH (1934) Hobart zoological collection as unique as marsupial wolf. *The Australasian*, Melbourne, 20 January.

Fleay DH (1937) Features of native animals. *Table Talk*, Melbourne, 1 October.

Fleay DH (1940) Bush Notes. *The Australasian*, Melbourne, 30 March.

Fleay-Thompson R (2007) *Animals First: The Story of Pioneer Australian Conservationist & Zoologist Dr David Fleay.* Keeaira Press, Southport.

Flies AS, Flies EJ, Fox S, Gilbert A, Johnson SR, *et al.* (2020) An oral bait vaccination approach for the Tasmanian devil facial tumor diseases. *Expert Review of Vaccines* **19**, 1–10. doi:10.1080/14760584.2020.1711058

Flower WH (1867) On the development and succession of the teeth in the Marsupialia. *Philosophical Transactions of the Royal Society of London* **157**, 631–641. doi:10.1098/rstl.1867.0020

Forth JS [Pseudonym for James Smith] (1862) Tasmanian tigers. *Launceston Examiner*, Launceston, 22 November.

Franklin TW, Mckelvey KS, Golding JD, Mason DH, Dysthe JC, *et al.* (2019) Using environmental DNA methods to improve winter surveys for rare carnivores: DNA from snow and improved noninvasive techniques. *Biological Conservation* **229**, 50–58. doi:10.1016/j.biocon.2018.11.006

Freeman CJ (2010) *Paper Tiger: A Visual History of the Thylacine.* Brill Academic Publishers, Leiden, The Netherlands.

Freeman CJ (2014) *Paper Tiger: How Pictures Shaped the Thylacine.* Forty South Publishing, Hobart.

Friedel E (1890) Zoobiologisches aus Paris. In *Der Zoologische Garten.* (Ed. FC Noll) pp. 245–251. Verlag von Mahlau & Waldschmidt, Frankfurt.

Fuller E (2013) *Lost Animals: Extinction and the Photographic Record.* Princeton University Press, Princeton.

Furió M, Ruiz-Sánchez FJ, Crespo VD, Freudenthal M, Montoya P (2012) The southernmost Miocene occurrence of the last European herpetotheriid *Amphiperatherium frequens* (Metatheria, Mammalia). *Comptes Rendus Palevol* **11**, 371–377. doi:10.1016/j.crpv.2012.01.004

Furlan EM, Gleeson D, Wisniewski C, Yick J, Duncan RP (2019) eDNA surveys to detect species at very low densities: a case study of European carp eradication in Tasmania, Australia. *Journal of Applied Ecology* **56**, 2505–2517. doi:10.1111/1365-2664.13485

Gaffney D, Summerhayes GR, Luu S, Menzies J, Douglas K, *et al.* (2021) Small game hunting in montane rainforests: specialised capture and broad spectrum foraging in the Late Pleistocene to Holocene New Guinea Highlands. *Quaternary Science Review* **252**, 106742. doi:10.1016/j.quascirev.2020.106742

Galan M, Pons J-B, Tournayre O, Pierre É, Leuchtmann M, *et al.* (2018) Metabarcoding for the parallel identification of several hundred predators and their prey: application to bat species diet analysis. *Molecular Ecology Resources* **18**, 474–489. doi: 10.1111/1755-0998.12749

Garde M (1997) 'Bawinanga Rock Art Recording Project: 1997 Field Season Report (MS3700)'. Maningrida Arts and Culture, Maningrida, NT.

Gelfo JN, Goin FJ, Bauzá N, Reguero MA (2019) The fossil record of Antarctic land mammals: commented review and hypotheses for future research. *Advances in Polar Science* **30**, 274–292. doi:10.13679/j.advps.2019.0021

Gemmell RT (1986) Lung development in the marsupial bandicoot, *Isoodon macrourus. Journal of Anatomy* **148**, 193–204.

Gillies SD, Morrison SL, Oi VT, Tonegawa S (1983) A tissue-specific transcription enhancer element is located in the major intron of a rearranged immunoglobulin heavy chain gene. *Cell* **33**, 717–728. doi:10.1016/0092-8674(83)90014-4

Glaskin K (2021) Extinction, inscription and the Dreaming: exploring a Thylacine connection. *Anthropological Forum* **31**, 165–185. doi:10.1080/00664677.2021.1937513

Godfrey M (1984) *Waratah Pioneer of the West.* Municipality of Waratah, Waratah.

Goin FJ, Woodburne MO, Zimicz AN, Martin GM, Chornogubsky L (2016) *A Brief History of South American Metatherians: Evolutionary Contexts and Intercontinental Dispersals.* Springer, New York.

Goldberg CS, Turner CR, Deiner K, Klymus KE, Thomsen PF, *et al.* (2016) Critical considerations for the application of environmental DNA methods to detect aquatic species. *Methods in Ecology and Evolution* **7**, 1299–1307. doi:10.1111/2041-210X.12595

Goldfuss GA (1820) *Handbuch der Zoologie. Dritter Theil, zweite Abtheilung.* Johann Leonhard Schrag, Nürnberg, Germany.

Gossage E (1981) Thylacine Competition entry NS896/1/6. Libraries Tasmania, Hobart.

Gould D (no date) Files held by the Enkler family, Perth.

Green RH (1967) Notes on the devil (*Sarcophilus harrisii*) and the quoll (*Dasyurus viverrinus*) in north-eastem Tasmania. *Records of the Queen Victoria Museum* **27**, 1–13.

Greentree C, Saunders G, Mcleod L, Hone J (2000) Lamb predation and fox control in south-eastern Australia. *Journal of Applied Ecology* **37**, 935–943. doi:10.1046/j.1365-2664.2000.00530.x.

Gregory GG, Munday BL, Beveridge I, Rickard MD (1975) Studies on *Anoplotaenia dasyuri* Beddard, 1911 (Cestoda: Taeniidae), a parasite of the Tasmanian devil: life-cycle and epidemiology. *International Journal for Parasitology* **5**, 187–191. doi:10.1016/0020-7519(75)90027-2

Gregory RL (1970) Techniques and apparatus for the study of visual perception. *Scientific Progress* **58**, 357-378.

Griffith OW, Chavan AR, Pavlicev M, Protopapas S, Callahan R, *et al.* (2019) Endometrial recognition of pregnancy occurs in the grey short-tailed opossum (*Monodelphis domestica*). *Proceedings of the Royal Society B: Biological Sciences* **286**, 20190691. doi:10.1098/rspb.2019.0691

Griffiths J (1972) The search for the Tasmanian tiger. *Natural History* **10**, 70–88.

Grose M, Hohnen M (2014) *Thylacine & Red Kangaroo,* motion picture. Directed by P Williams. Skinnyfish Music.

Grueber CE, Chong R, Gooley RM, McLenna EA, Barrs VR, *et al.* (2020) Genetic analysis of scat samples to inform conservation of Tasmanian devil. *Australian Zoologist* **40**, 492–504. doi:10.7882/AZ.2020.005

Guiler ER (unpublished) Thylacine Bounty: details of claims compiled from Lands Department accounts for years 1888–1912. Hobart, 1958.

Guiler ER (1958) The Thylacine. *Australian Museum Magazine* **12**, 352–354.

Guiler ER (1961) The former distribution and decline of the thylacine. *Australian Journal of Science* **23**, 207–210.

Guiler ER (1978) Observations on the Tasmanian devil, *Sarcophilus harrisii* (Dasyuridae: Marsupialia) at Granville Harbour, 1966–75. *Papers and Proceedings of the Royal Society of Tasmania* **112**, 161–188. doi:10.26749/rstpp.112.161

Guiler ER (1985) *Thylacine: The Tragedy of the Tasmanian Tiger*. Oxford University Press, Melbourne.

Guiler ER (1986) The Beaumaris Zoo in Hobart. *Papers and Proceedings of the Tasmanian Historical Research Association* **33**, 121–171.

Guiler ER (1991) *The Tasmanian Tiger in Pictures.* St David's Park Publishing, Hobart.

Guiler ER, Godard P (1998) *Tasmanian Tiger: A Lesson to Be Learnt.* Abrolhos Publishing, Perth.

Gunn RC (1838) Notices accompanying a collection of quadrupeds and fish from Van Diemen's Land. *Annals and Magazine of Natural History: Zoology, Botany and Geology* **1**, 101–111. doi:10.1080/00222933809512244

Hamede R, Jones ME, McCallum H (2012) Biting injuries and transmission of Tasmanian devil facial tumour disease. *Journal of Animal Ecology* **82**, 182–190. doi:10.1111/j.1365-2656.2012.02025.x

Hamer RP, Gardiner RZ, Johnson CN, Jones ME (2021) A triple threat: high population density, high foraging intensity and flexible habitat preferences explain high impact of feral cats on prey. *Proceedings of the Royal Society B: Biological Sciences* **288**, 20201194. doi:10.1098/rspb.2020.1194

Hansen BD, Harley DKP, Lindenmayer DB, Taylor AC (2009) Population genetic analysis reveals long-term decline of a threatened endemic Australian marsupial. *Molecular Ecology* **18**, 3346–3362. doi:10.1111/j.1365-294X.2009.04269.x

Harris GP (1808) Description of two new species of *Didelphis* from Van Diemen's Land. *Transactions of the Linnean Society of London* **9**, 174–178. doi:10.1111/j.1096-3642.1818.tb00336.x

Harrison J (unpublished) Notebook entry, author's copy, 1928.

Haygarth N (2016) Clem Penney's stripey doormat. *Nic Haygarth.* <https://nichaygarth.com/index.php/2016/11/27/clem-penneys-stripy-doormat/>.

Haygarth N (2017) The myth of the dedicated thylacine hunter: stockman-hunter culture and the decline of the thylacine (Tasmanian tiger) in Tasmania during the late nineteenth and early twentieth centuries. *Papers and Proceedings: Tasmanian Historical Research Association* **33**, 30–35. doi:10.3316/informit.047041010462417

Haynes J (2001) The marsupial and monotreme thymus, revisited. *Journal of Zoology* **253**, 167–173. doi:10.1017/S0952836901000152

HCCRC (1923a) Minutes of the Hobart City Council Reserves Committee MCC16/72/6. Libraries Tasmania, Hobart, 5 June.

HCCRC (1923b) Minutes of the Hobart City Council Reserves Committee MCC16/72/6. Libraries Tasmania, Hobart, 11 September.

HCCRC (1924a) Minutes of the Hobart City Council Reserves Committee MCC16/72/7. Libraries Tasmania, Hobart, 19 February.

HCCRC (1924b) Minutes of the Hobart City Council Reserves Committee MCC16/72/4. Libraries Tasmania, Hobart, 19 February.

HCCRC (1925a) Minutes of the Hobart City Council Reserves Committee MCC16/72/8. Libraries Tasmania, Hobart, 5 May.

HCCRC (1925b) Minutes of the Hobart City Council Reserves Committee MCC16/72/8. Libraries Tasmania, Hobart, 18 August

HCCRC (1928a) Minutes of the Hobart City Council Reserves Committee MCC16/72/10. Libraries Tasmania, Hobart, 27 March.

HCCRC (1928b) Minutes of the Hobart City Council Reserves Committee MCC16/72/10. Libraries Tasmania, Hobart, 17 April.

HCCRC (1930) Minutes of the Hobart City Council Reserves Committee MCC16/72/13. Libraries Tasmania, Hobart, 30 October.

HCCRC (1936a) Minutes of the Hobart City Council Reserves Committee MC-16-72-17. Libraries Tasmania, Hobart, 16 September.

HCCRC (1936b) Minutes of the Hobart City Council Reserves Committee MC-16-72-17. Libraries Tasmania, Hobart, 19 February.

HCCRC (1937) Minutes of the Hobart City Council Reserves Committee MC-16-72-17. Libraries Tasmania, Hobart, 16 June.

Heck L (1912) *Die Säugetiere von Alfred Brehm, Eriter Band*. Bibliographisches Institut, Leipzig.

Helgen KM, Veatch EG (2015) Recently extinct Australian marsupials and monotremes. In *Handbook of the Mammals of the World. Vol. 5. Monotremes and Marsupials*. (Eds DE Wilson and RA Mittermeier) pp. 17–31. Lynx Edicions, Barcelona.

Hoare P (2014) *The Sea Inside*. Melville House, London.

Hobart Town Gazette (1826) Hobart, 2 December.

Hobart Town Gazette (1921) Cited in Sleightholme SR, Campbell RC (2016) A retrospective assessment of 20th century thylacine populations. *Australian Zoologist* **38**, 102–129. doi:10.7882/AZ.2015.023

Hobart Town Gazette and Van Diemen's Land Advertiser (1821) Hobart, 3 November.

Hobart Town Gazette and Van Diemen's Land Advertiser (1823) Hobart, 2 August.

Hogg C, Fox S, Pemberton D, Belov K (Eds) (2019) *Saving the Tasmanian Devil: Recovery through Science-based Management*. CSIRO Publishing, Melbourne.

Horovitz I, Sánchez-Villagra MR (2003) A morphological analysis of marsupial mammal higher-level phylogenetic relationships. *Cladistics* **19**, 181–212. doi:10.1111/j.1096-0031.2003.tb00363.x

Horton R, Wilming L, Rand V, Lovering RC, Bruford EA, *et al.* (2004) Gene map of the extended human MHC. *Nature Reviews Genetics* **5**, 889–899. doi:10.1038/nrg1489

Huet J (1887) Le Thylacine a tête de chien. *Le Naturaliste revue illustrée des Sciences Naturelles* **1**, 68–70.

Illustrated Tasmanian Mail (1934) I Remember. Hobart, 13 December.

Ingram J, Kirkpatrick JB (2013) Native vertebrate herbivores facilitate native plant dominance in old fields while preventing native tree invasion: implications for threatened species. *Pacific Conservation Biology* **19**, 331–342. doi:10.1071/PC130331

Jackson S (1937) The Tasmanian tiger. *The World's News*, Sydney, 5 May.

Jackson S, Groves C (2015) *Taxonomy of Australian Mammals*. CSIRO Publishing, Melbourne.

Janis CM, Figueirido B (2014) Forelimb anatomy and the discrimination of the predatory behavior of carnivorous mammals: the thylacine as a case study. *Journal of Morphology* **275**, 1321–1338. doi:10.1002/jmor.20303

Jenkins K (1981) Thylacine Competition entry NS896/4. Libraries Tasmania, Hobart.

Johnson CN, Wroe S (2003) Causes of extinction of vertebrates during the Holocene of mainland Australia: arrival of the dingo, or human impact? *Holocene* **13**, 941–948. doi:10.1191/0959683603hl682fa

Johnson K, Roff A (1982) The Western Quoll, *Dasyurus geoffroil* (Dasyuridae, Marsupalia) in the Northern Territory: historical records from venerable sources. In *Carnivorous marsupials.* (Ed. M Archer) pp. 221–226. Royal Zoological Society of New South Wales, Sydney.

Johnstone J (1898) The thymus in marsupials. *Journal of the Linnean Society of London, Zoology* **25**, 537–557.

Jones ME (1995) Guild structure of the large marsupial carnivores in Tasmania. PhD thesis. The University of Tasmania, Australia.

Jones ME (1997) Character displacement in Australian dasyurid carnivores: size relationships and prey size patterns. *Ecology* **78**, 2569–2587. doi:10.1890/0012-9658(1997)078[2569:CDIADC]2.0.CO;2

Jones ME (2003) Convergence in ecomorphology and guild structure among marsupial and placental carnivores. In *Predators with Pouches: The Biology of Carnivorous Marsupials.* (Eds M Jones, C Dickman and M Archer) pp. 285–296. CSIRO Publishing, Melbourne.

Jones ME, Hamede RH, Hollings T, McCallum H (2019) Devils and disease in the landscape: the impact of disease on devils in the wild and on the Tasmanian ecosystem. In *Saving the Tasmanian Devil: Recovery Using Science-based Management.* (Eds C Hogg, S Fox, D Pemberton and K Belov) pp. 85–100. CSIRO Publishing, Melbourne.

Jones ME, Stoddart DM (1998) Reconstruction of the predatory behaviour of the extinct marsupial thylacine (*Thylacinus cynocephalus*). *Journal of Zoology* **246**, 239–246. doi:10.1111/j.1469-7998.1998.tb00152.x

Jordan AM (1987) *The Tiger Man, The Thylacine: Yesterday, Today, and Tomorrow; My Study and Findings*, Wordswork Express, Perth.

Kamiya T, Dwyer KO, Nakagawa S, Poulin R (2014) What determines species richness of parasitic organisms? A meta-analysis across animal, plant and fungal hosts. *Biological Reviews* **89**, 123–134. doi:10.1111/brv.12046

Karadada J, Karadada L, Goonack W, Mangolamara G, Bunjuck W, *et al.* (2011) *Uunguu Plants and Animals.* Wunambal Gaambera Aboriginal Corporation/Department of Natural Resources, Environment, The Arts and Sport, Kalumburu.

Karlen SJ, Krubitzer L (2007) The functional and anatomical organization of marsupial neocortex: evidence for parallel evolution across mammals. *Progress in Neurobiology* **82**, 122–141. doi:10.1016/j.pneurobio.2007.03.003

Kealy S, Beck R (2017) Total evidence phylogeny and evolutionary timescale for Australian faunivorous marsupials (Dasyuromorphia). *BMC Evolutionary Biology* **17**, 240. doi:10.1186/s12862-017-1090-0

Kerr IAR, Prideaux GJ (2022) A new genus of kangaroo (Marsupialia, Macropodidae) from the late Pleistocene of Papua New Guinea. *Transactions of the Royal Society of South Australia.* doi:10.1080/03721426.2022.2086518

Kimber D (2011) *Cultural Values Associated with Alice Springs Water*. Department of Environment, Parks and Water Security, NT, <https://denr.nt.gov.au/__data/assets/pdf_file/0005/254606/Cultural-Values-of-Alice-Water-Kimber-2011.pdf>

Kirkpatrick JB, Bridle KL (Eds) (2007) *People, Sheep and Nature Conservation: The Tasmanian Experience.* CSIRO Publishing, Melbourne.

Knights T, Langley MC (2021) Invisible or ignored: investigating the lack of Thylacine-based material culture in the Australian archaeological record. *Archaeology in Oceania* **56**, 100–110. doi:10.1002/arco.5227

Kohl P (2017) Reclaiming hope in extinction storytelling. In *Recreating the wild: de-extinction, technology, and the ethics of conservation*, special report. *Hastings Center Report* **47**, S24–S29. doi:10.1002/hast.748

Krefft G (1868) Description of a new species of thylacine (*Thylacinus breviceps*). *The Annals and Magazine of Natural History; Zoology, Botany, and Geology* **2**, 296–297.

Krubitzer L (2007) The magnificent compromise: cortical field evolution in mammals. *Neuron* **56**, 201–208. doi:10.1016/j.neuron.2007.10.002

Kwak ML (2018) Australia's vanishing fleas (Insecta: Siphonaptera): a case study in methods for the assessment and conservation of threatened flea species. *Journal of Insect Conservation* **22**, 545–550. doi:10.1007/s10841-018-0083-7

Kwak ML, Madden C, Wicker L (2017) The first record of the native flea *Acanthopsylla rothschildi* Rainbow, 1905 (Siphonaptera: Pygiopsyllidae) from the endangered Tasmanian devil (*Sarcophilus harrisii* boitard, 1841), with a review of the fleas associated with the Tasmanian devil. *The Australian Entomologist* **44**, 293–296. doi:10.3316/INFORMIT.320051305337564

Kwon YM, Gori K, Park N, Potts N, Swift K, *et al.* (2020) Evolution and lineage dynamics of a transmissible cancer in Tasmanian devils. *PLoS Biology* **18**, e3000926. doi:10.1371/journal.pbio.3000926

Ladle RJ, Jepson P, Araújo MB, Whittaker RJ (2004) Dangers of crying wolf over risk of extinctions. *Nature* **428**, 799. doi:10.1038/428799b

Laird N (1968) Keeping Tasmanian tiger in captivity would be an outrage. *The Mercury*, Hobart, 7 October.

Laird N (no date) Note NG1143 held by Libraries Tasmania, Hobart.

Lansley L (1978) Interview with Erik Bierre ID: 2080938. National Library of Australia, Canberra.

Launceston Advertiser (1836) Launceston, 23 June.

Launceston Courier (1840) Launceston, 26 October.

Launceston Examiner (1879) Notices to correspondents. Launceston, 26 November.

Launceston Examiner (1883a) Tasmanian Tigers. Launceston, 24 August.

Launceston Examiner (1883b) Tasmanian Tigers. Launceston, 7 September.

Launceston Examiner (1884) Tasmanian Tigers. Launceston, 5 January.

Launceston Examiner (1897) The Tasmanian tiger. Launceston, 10 July.

Le Souëf WHD (1907) *Wildlife in Australia.* Whitcombe and Tombs, Melbourne.

Letnic M, Fillios M, Crowther MS (2012) Could direct killing by larger dingoes have caused the extinction of the thylacine from mainland Australia? *PLoS ONE* **7**, e34877. doi:10.1371/journal.pone.0034877

Letnic M, Story P, Story G, Field J, Brown O, et al. (2011) Resource pulses, switching trophic control, and the dynamics of small mammal assemblages in arid Australia. *Journal of Mammalogy* **92**, 1210–1222. doi:10.1644/10-MAMM-S-229.1

Lewis D (2017) Megafauna identification for dummies: Arnhem Land and Kimberley 'megafauna' paintings. *Rock Art Research* **34**, 82–99.

Linnard G, Williams M, Holmes B (2020) The parson, the psychiatrist, the publican and his nephews: the two final thylacine captures in Tasmania. *Papers and Proceedings: Tasmanian Historical Research Association* **67**, 6–22. doi:10.3316/INFORMIT.586826815432203

Llamas B, Valverde G, Fehren-Schmitz L, Weyrich LS, Cooper A, *et al.* (2017) From the field to the laboratory: controlling DNA contamination in human ancient DNA research in the high-throughput sequencing era. *STAR: Science & Technology of Archaeological Research* **3**, 1–14. doi:10.1080/20548923.2016.1258824

Long JA (2002) Dasyures, numbats and thylacines: diverse eaters of flesh. In *Prehistoric Mammals of Australia and New Guinea: One Hundred Million Years of Evolution.* (Eds JA Long, M Archer, T Flannery and S Hand) pp. 85–100. UNSW Press, Sydney.

Lord C (unpublished a) 'Export of native Australian animals and birds – including thylacine' P437, National Archive of Australia, Canberra, 20 July.

Lord C (unpublished b) Letter to Chairman of Fauna Board, author's copy, 27 March 1929.

Lord C (unpublished c) Memo to superintendents, Libraries Tasmania, Hobart, 24 March 1937.

Lourandos H (1997) *Continent of Hunter-Gatherers: New Perspectives in Australian Prehistory.* Cambridge University Press, Melbourne.

Lowry DC, Lowry WJ (1967) Discovery of a Thylacine (Tasmanian Tiger) carcase in a cave near Eucla, Western Australia. *Helictite* **5**, 25–29.

Luckett WP (1993) Ontogenetic staging of the mammalian dentition, and its value for assessment of homology and heterochrony. *Journal of Mammalian Evolution* **1**, 269–282. doi:10.1007/BF01041667

Luckett WP, Hong Luckett N, Harper T (2019) Microscopic analysis of the developing dentition in the pouch young of the extinct marsupial *Thylacinus cynocephalus*, with an assessment of other developmental stages and eruption. *Memoirs of Museum Victoria* **78**, 1–21. doi:10.24199/j.mmv.2019.78.01

Luckett WP, Woolley PA (1996) Ontogeny and homology of the dentition in dasyurid marsupials: development in *Sminthopsis virginiae*. *Journal of Mammalian Evolution* **3**, 327–364.

Lyne AG, Verhagen AMW (1957) Growth of the marsupial *Trichosurus vulpecula* and a comparison with some higher mammals. *Growth* **21**, 167–195.

MacFarlane AM, Coulson G, Ruckstuhl K, Neuhaus P (2005) Sexual segregation in Australian marsupials. In *Sexual Segregation in Vertebrates*. (Eds K Ruckstuhl and P Neuhaus) pp. 254–279. Cambridge University Press, Cambridge, UK.

Macintosh NWG (1971) Analysis of an Aboriginal skeleton and a pierced tooth necklace from Lake Nitchie, Australia. *Anthropologie* **9**, 49–62.

MacKenzie C (1924) The medical importance of the native animals of Australia. Paper circulated for the information of Honourable Members by the Honourable the Chief Secretary. MacKenzie papers, National Museum of Australia.

Mangglamarra G, Burbidge AA, Fuller PJ (1991) Wunambal words for rainforest and other Kimberley plants and animals. In *Kimberley Rainforests of Australia*. (Eds NL McKenzie, RB Johnston and PG Kendrick) pp. 413–421. Surrey Beatty and Sons, Sydney.

Maynard D, Gordon T (2014) *Tasmanian Tiger: Precious Little Remains*. Queen Victoria Museum & Art Gallery, Launceston.

McCallum H, Dobson A (1995) Detecting disease and parasite threats to endangered species and ecosystems. *Trends in Ecology & Evolution* **10**, 190–194. doi:10.1016/s0169-5347(00)89050-3

McCallum H, Jones M (2006) To lose both would look like carelessness: Tasmanian devil facial tumour disease. *PLoS Biology* **4**, e342. doi:10.1371/journal.pbio.0040342

McCallum H, Jones M, Hawkins C, Hamede R, Lachish S, *et al.* (2009) Transmission dynamics of Tasmanian devil facial tumor disease may lead to disease-induced extinction. *Ecology* **90**, 3379–3392. doi:10.1890/08-1763.1

McKenzie NL, Johnston RB, Kendrick PG (Eds) (1991) *Kimberley Rainforests of Australia*. Surrey Beatty and Sons, Sydney.

Meek PD, Ballard G-A, Vernes K, Fleming PJS (2015) The history of wildlife camera trapping as a survey tool in Australia. *Australian Mammalogy* **37**, 1–12. doi:10.1071/AM14021

Mendel LC, Kirkpatrick JB (2002) Historical progress of biodiversity conservation in the protected-area system of Tasmania, Australia. *Conservation Biology* **16**, 1–11.

Menzies BR, Renfree MB, Heider T, Mayer F, Hildebrandt TB, *et al.* (2012) Limited genetic diversity preceded extinction of the Tasmanian tiger. *PLoS ONE* **7**, e35433. doi:10.1371/journal.pone.0035433

Milligan J (1853) Remarks upon the Habits of the Wombat, the Hyaena, and Certain Reptiles. *Papers and Proceedings of the Royal Society of Van Dieman's Land* **2**, 310.

Milne-Edwards A (1891) Muséum d'histoire naturelle. La Ménagerie (Rapport au Ministre de l'Instruction publique). G. Masson, Paris.

Mitchel J, Sargent R (1996) Oral history interview: Alison Reid, 14 January. Libraries Tasmania. <https://www.youtube.com/watch?v=MAGRCnZ4K10>.

Mittelbach M, Crewdson M (2005) *Carnivorous Nights: On the Trail of the Tasmanian Tiger*. Villard, New York.

Moeller H (1968) Zur Frage der Parallelerscheinungen bei Metatheria und Eutheria. Vergleichende Untersuchungen an Beutelwolf und Wolf. *Zeitschrift für Wissenschaftliche Zoologie* **177**, 283–392.

Moeller HF (1997) *Der Beutelwolf. Thylacinus cynocephalus.* Westarp Wissenschaften, Magdeburgh, Germany.

Mooney N (2014) So close and yet so far. In *The Tasmanian Tiger: Extinct or Extant?* (Ed. R Lang) pp. 37–49. Strange Nation Publishing, Hazelbrook, New South Wales.

Morris B, Cheng Y, Warren W, Papenfuss AT, Belov K (2015) Identification and analysis of divergent immune gene families within the Tasmanian devil genome. *BMC Genomics* **16**, 1017. doi:10.1186/s12864-015-2206-9

Morris K, Austin JJ, Belov K (2013) Low major histocompatibility complex diversity in the Tasmanian devil predates European settlement and may explain susceptibility to disease epidemics. *Biology Letters* **9**, 20120900. doi:10.1098/rsbl.2012.0900

Mountain M-J (1991) Highland New Guinea hunter-gatherers: the evidence of Nombe Rockshelter, Simbu, with emphasis on the pleistocene. PhD thesis (2 vols). Australian National University, Canberra.

Mudway M (2019) *Stripes Ascending.* Self-Published, Ontario.

Muirhead J (1992) A specialised thylacinid, *Thylacinus macknessi*, (Marsupialia: Thylacinidae) from Miocene deposits of Riversleigh, northwestern Queensland. *Australian Mammalogy* **15**, 67–76.

Mulvaney KJ (2019) 'Murujuga Petroglyphs: Rock art narratives'. Paper presented at the University of Western Australia, Crawley, Perth, WA, 16 May.

Murchison EP, Tovar C, Hsu A, Bender HS, Kheradpour P, *et al.* (2010) The Tasmanian devil transcriptome reveals Schwann cell origins of a clonally transmissible cancer. *Science* **327**, 84–87. doi:10.1126/science.1180616

Murray (1925) For Sale, Tasmanian tiger. *The Mercury*, Hobart, 13, 15, 17 June.

Murray P, Megirian D (2000) Two new genera and three new species of Thylacinidae (Marsupialia) from the Miocene of the Northern Territory, Australia. *The Beagle: Records of the Museums and Art Galleries of the Northern Territory* **16**, 145–162. doi:10.5962/p.254543

Murray PF, Megirian D (2006) The Pwerte Marnte Marnte Local Fauna: a new vertebrate assemblage of presumed Oligocene age from the Northern Territory of Australia. *Alcheringa: An Australasian Journal of Palaeontology* **30**, 211–228. doi:10.1080/03115510609506864

Nabarlambarl SP, Nabarlambarl ED, Dangbunala B (2018) *Djankerrk. Thylacine Story.* Warddeken Land Management, Darwin.

Newton AH, Spoutil F, Prochazka J, Black JR, Medlock K, *et al.* (2018) Letting the "cat" out of the bag: pouch young development of the extinct Tasmanian tiger revealed by X-ray computed tomography. *Royal Society Open Science* **5**, 171914. doi:10.1098/rsos.171914

Newton AH, Weisbecker V, Pask AJ, Hipsley CA (2021) Ontogenetic origins of cranial convergence between the extinct marsupial thylacine and placental gray wolf. *Nature Communications Biology* **4**, 51. doi:10.1038/s42003-020-01569-x

Nghiem LT, Papworth SK, Lim FK, Carrasco LR (2016) Analysis of the capacity of Google Trends to measure interest in conservation topics and the role of online news. *PloS ONE* **11**, e0152802. doi:10.1371/journal.pone.0152802

Obendorf DL, Smith SJ (1989) A re-examination of *Dithyridium cynocephali* Ransom 1905, a metacestode parasite from the thylacine *Thylacinus cynocephalus. Papers and Proceedings of the Royal Society of Tasmania* **123**, 133–136. doi:10.26749/rstpp.123.133

Old JM (2015) Immunological insights into the life and times of the extinct Tasmanian tiger (*Thylacinus cynocephalus*). *PLoS ONE* **10**, e0144091. doi:10.1371/journal.pone.0144091

Old JM (2016) Haematopoiesis in marsupials. *Developmental and Comparative Immunology* **58**, 40–46. doi:10.1016/j.dci.2015.11.009

Old JM, Deane EM (2000) Development of the immune system and immunological protection in marsupial pouch young. *Developmental and Comparative Immunology* **24**, 445–454. doi:10.1016/S0145

Owen D, Pemberton D (2005) *Tasmanian Devil: A Unique and Threatened Animal.* Allen & Unwin, Sydney.

Owen R (1845) *Descriptive and Illustrated Catalogue of the Fossil Organic Remains of Mammalia and Aves Contained in the Museum of the Royal College of Surgeons of England.* Richard and John E. Taylor, London.

Paddle RN (1992) Last resting place of a thylacine. *Nature* **360**, 215. doi:10.1038/360215a0

Paddle RN (1996) Mueller's magpies and marsupial wolves: a window into "What might have been". *The Victorian Naturalist* **113**, 215–217.

Paddle RN (2000) *The Last Tasmanian Tiger: The History and Extinction of the Thylacine.* Cambridge University Press, Melbourne.

Paddle RN (2008) The most photographed of thylacines: Mary Roberts' Tyenna male – including a response to Freeman (2005) and a farewell to Laird (1968). *Australian Zoologist* **4**, 459–470. doi:10.7882/AZ.2008.024

Paddle RN (2012) The thylacine's last straw: epidemic disease in a recent mammalian extinction. *Australian Zoologist* **36**, 75–92. doi:10.7882/AZ.2012.008

Panasci M, Ballard WB, Breck S, Rodriguez D, Densmore LD, *et al.* (2011) Evaluation of fecal DNA preservation techniques and effects of sample age and diet on genotyping success. *The Journal of Wildlife Management* **75**, 1616–1624. doi:10.1002/jwmg.221

Patchett AL, Coorens THH, Darby J, Wilson R, McKay MJ, *et al.* (2019) Two of a kind: transmissible Schwann cell cancers in the endangered Tasmanian devil (*Sarcophilus harrisii*). *Cellular and Molecular Life Sciences* **77**, 1847–1858. doi:10.1007/s00018-019-03259-2

Patchett AL, Flies AS, Lyons AB, Woods GM (2020) Curse of the devil: molecular insights into the emergence of transmissible cancers in the Tasmanian devil (*Sarcophilus harrisii*). *Cellular and Molecular Life Sciences* **77**, 2507–2525. doi:10.1007/s00018-019-03435-4

Paterson W (1805a) Private letter to Sir Joseph Banks, dated 30 March. Held by the State Library of New South Wales.

Paterson W (1805b) The Sydney Gazette and NSW Advertiser, Sydney, 21 April.

Patton AH, Margres MJ, Stahlke AR, Hendricks S, Lewallen K, *et al.* (2019) Contemporary demographic reconstruction methods are robust to genome assembly quality: a case study in Tasmanian Devils. *Molecular Biology and Evolution* **36**, 2906–2921. doi:10.1093/molbev/msz191

Patton AH, Lawrance MF, Margres MJ, Kozakiewicz CP, Hamede R, *et al.* (2020) A transmissible cancer shifts from emergence to endemism in Tasmanian devils. *Science* 370, eabb9772. doi:10.1126/science.abb9772

Paull J (2011) Environmental management in Tasmania: better off dead? In *Island Futures: Conservation and Development Across the Asia-Pacific Region.* (Eds G Baldacchino and N Daniels) pp. 153–168. Springer, Tokyo.

Pearse AM (1981) Aspects of the biology of *Uropsylla tasmanica* Rothschild (Siphonaptera). PhD thesis. The University of Tasmania, Australia.

Pearse AM, Swift K (2006) Allograft theory: transmission of devil facial-tumour disease. *Nature* **439**, 549. doi:10.1038/439549a

Peel E, Frankenberg SR, Hogg CJ, Pask A, Belov K (2021) Annotation of immune genes in the extinct thylacine (*Thylacinus cynocephalus*). *Immunogenetics* **73**, 263–275. doi:10.1007/s00251-020-01197-z

Pemberton D (1990) Social organisation and behaviour of the Tasmanian devil, *Sarcophilus harrisii.* PhD thesis. The University of Tasmania, Australia.

Pemberton D, Renouf, D (1993) A field study of communication and social behaviour of the Tasmanian devil at feeding sites. *Australian Journal of Zoology* **41**, 507–526. doi:10.1071/ZO9930507

Peregrine [Pseudonym for Michael Sharland] (1937) Moulting Birds. *The Mercury*, Hobart, 20 February.

Peregrine [Pseudonym for Michael Sharland] (1939) The Tasmanian tiger. *The Mercury*, Hobart, 25 March.

Peters KJ, Saltré F, Friedrich T, Jacobs Z, Wood R, *et al.* (2019) FosSahul 2.0, an updated database for the Late Quaternary fossil records of Sahul. *Scientific Data* **6**, 1–7. doi:10.1038/s41597-019-0267-3

Piper CJ, Veth PM (2021) Palaeoecology and sea level changes: decline of mammal species richness during late Quaternary island formation in the Montebello Islands, north-western Australia. *Palaeontologica Electronica* **24**, 1–21. doi:10.26879/1050

Pledge NS (1992) The Curramulka local fauna: a new late Tertiary fossil assemblage from Yorke Peninsula, South Australia. *The Beagle: Records of the Museums and Art Galleries of the Northern Territory* **9**, 115–142.

Prevosti FJ, Forasiepi A, Zimicz N (2013) The evolution of the Cenozoic terrestrial mammal guild in South America: competition or replacement? *Journal of Mammalian Evolution* **20**, 3–21. doi:10.1007/s10914-011-9175-9

Prideaux GJ, Kerr IAR, van Zoelen JD, Grun R, van der Kaars S, *et al.* (in press) Re-evaluating the evidence for late-surviving megafauna at Nombe rock shelter in the New Guinea highlands. *Archaeology in Oceania.*

Prowse TAA, Johnson CN, Lacy RC, Bradshaw CJA, Pollak JP, *et al.* (2013) No need for disease: testing extinction hypotheses for the thylacine using multi-species metamodels. *Journal of Animal Ecology* **82**, 355–364. doi:10.1111/1365-2656.12029

Prowse TAA, Johnson CN, Bradshaw CJA, Brook BW (2014) An ecological regime shift resulting from disrupted predator–prey interactions in Holocene Australia. *Ecology* **95**, 693–702. doi:10.1890/13-0746.1

Pye RJ, Hamede R, Siddle HV, Caldwell A, Knowles GW, *et al.* (2016a) Demonstration of immune responses against devil facial tumour disease in wild Tasmanian devils. *Biology Letters* **12**, 5. doi:10.1098/rsbl.2016.0553

Pye RJ, Pemberton D, Tovar C, Tubio JM, Dun KA, *et al.* (2016b) A second transmissible cancer in Tasmanian devils. *Proceedings of the National Academy of Sciences, USA* **113**, 374–378. doi:10.1073/pnas.1519691113

Pye RJ, Patchett A, McLennan E, Thomson R, Carver S, *et al.* (2018) Immunization strategies producing a humoral IgG immune response against Devil Facial Tumor Disease in the majority of Tasmanian devils destined for wild release. *Frontiers in Immunology* **9**, 259. doi:10.3389/fimmu.2018.00259

Ralls K (1977) Sexual dimorphism in mammals: avian models and unanswered questions. *American Naturalist* **111**, 917–938.

Ransom BH (1905) Tapeworm cysts (*Dithyridium cynocephali* n.sp.) in the muscles of the Tasmanian Wolf (*Thylacinus cynocephalus*). *Transactions of the American Microscopical Society* **27**, 31–32.

Renshaw G (1905) *More Natural History Essays*. Sherratt and Hughes, London.

Rovinsky DS, Evans AR, Adams JW (2019) The pre-Pleistocene fossil thylacinids (Dasyuromorphia: Thylacinidae) and the evolutionary context of the modern thylacine. *PeerJ* **7**, e7457. doi:10.7717/peerj.7457

Rovinsky DS, Evans AR, Martin DG, Adams JW (2020) Did the thylacine violate the costs of carnivory? Body mass and sexual dimorphism of an iconic Australian marsupial. *Proceedings of the Royal Society B: Biological Sciences* **287**, 20201537. doi:10.1098/rspb.2020.1537

Rowan W (unpublished) William Rowan Archives, University of Alberta Libraries, Canada.

Roycroft E, MacDonald AJ, Moritz C, Moussalli A, Portela Miguez R, *et al.* (2021) Museum genomics reveals the rapid decline and extinction of Australian rodents since European settlement. *PNAS* **119,** e2021390118. doi:10.1073/pnas.2021390118

Ruckstuhl KE, Neuhaus P (2002) Sexual segregation in ungulates: a comparative test of three hypotheses. *Biological Reviews* **77**, 7796. doi:10.1017/s1464793101005814

Ruiz-Aravena M, Jones ME, Carver S, Estay S, Espejo C, *et al.* (2018) Sex bias in ability to cope with cancer: Tasmanian devils and facial tumour disease. *Proceedings of the Royal Society B: Biological Sciences* **285**, 20182239. doi:10.1098/rspb.2018.2239

Runciman SIC, Baudinette RV, Gannon BJ (1996) Postnatal development of the lung parenchyma in a marsupial: The Tammar wallaby. *The Anatomical Record* **244**, 193–206. doi:10.1002/(SICI)1097-0185(199602)244:2<193::AID-AR7>3.0.CO;2-2

Saint-Hilaire ÉG (1810) Description de deux espèces de Dasyures (*Dasyrus [sic] cynocephalus et Dasyurus ursinus*). *Annales du Muséum d'Histoire Naturelle, Paris* **15**, 301–306.

Sales NG, Kaizer M Da C, Coscia I, Perkins JC, Highlands A, *et al.* (2020) Assessing the potential of environmental DNA metabarcoding for monitoring Neotropical mammals: a case study in the Amazon and Atlantic Forest, Brazil. *Mammal Review* **50**, 221–225. doi:10.1111/mam.12183

Saltré F, Brook BW, Rodríguez-Rey M, Cooper A, Johnson CN, *et al.* (2015) Uncertainties in dating constrain model choice for inferring extinction time from fossil records. *Quaternary Science Reviews* **112**, 128–137. doi:10.1016/j.quascirev.2015.01.022

Schnell IB, Sollmann R, Calvignac-Spencer S, Siddall ME, Yu DW, *et al.* (2015) iDNA from terrestrial haematophagous leeches as a wildlife surveying and monitoring tool: prospects, pitfalls and avenues to be developed. *Frontiers in Zoology* **12**, 1–14. doi:10.1186/s12983-015-0115-z

Scoleri V (2020) Conservation introduction of top predator to an island triggers ecological cascades. PhD thesis. University of Tasmania, Hobart.

Serena M, Soderquist T (1989) Spatial organization of a riparian population of the carnivorous marsupial *Dasyurus geoffroii*. *Journal of Zoology* **219**, 373–383. doi:10.1111/j.1469-7998.1989.tb02586.x

Shapiro (2008) Two Day Estate Auction. *The Australian Jewish News*, Sydney, 12 September.

Shapiro B (2017) Pathways to de-extinction: how close can we get to resurrection of an extinct species? *Functional Ecology* **31**, 996–1002. doi:10.1111/1365-2435.12705

Sharland MSR (1923) The Beaumaris Zoo. *The Mercury*, Hobart, 20 June.

Sharland MSR (1939) In search of the Thylacine. *Papers and Proceedings of the Royal Zoological Society of New South Wales* **59**, 1–38.

Sharland MSR (1957) In search of the vanished "tiger". *People*, Sydney, 3 April.

Sharland MSR (1962) *Tasmanian Wildlife: A Popular Account of the Furred Land Mammals, Snakes and Introduced Mammals of Tasmania.* Melbourne University Press, Melbourne.

Sharland MSR (1972) Cited in Paddle R (2000) *The Last Tasmanian Tiger: The History and Extinction of the Thylacine.* p. 190. Cambridge University Press, Cambridge.

Siddle HV, Kreiss A, Eldridge MD, Noonan E, Clarke CJ, *et al.* (2007) Transmission of a fatal clonal tumor by biting occurs due to depleted MHC diversity in a threatened carnivorous marsupial. *Proceedings of the National Academy of Sciences, USA* **104**, 16221–16226. doi:10.1073/pnas.0704580104

Siddle HV, Kreiss A, Tovar C, Yuen CK, Cheng YY, *et al.* (2013) Reversible epigenetic down-regulation of MHC molecules by devil facial tumour disease illustrates immune escape by a contagious cancer. *Proceedings of the National Academy of Sciences, USA* **110**, 5103–5108. doi:10.1073/pnas.1219920110

Signor PW, Lipps JH (1982) Sampling bias, gradual extinction patterns and catastrophes in the fossil record. *Geological Society of America Special Papers* **190**, 291–296. doi:10.1130/SPE190-p291

Sleightholme SR, Ayliffe N (2017) *International Thylacine Specimen Database* (6th revision). [DVD-ROM: Master Copy] Zoological Society of London, London.

Sleightholme SR, Campbell CR (2016) A retrospective assessment of 20th century thylacine populations. *Australian Zoologist* **38**, 102–129. doi:10.7882/AZ.2015.023

Sleightholme SR, Campbell CR (2018) The International Thylacine Specimen Database (6th Revision: Project Summary & Final Report). *Australian Zoologist* **39**, 480–512. doi:10.7882/AZ.2017.011

Sleightholme SR, Campbell CR (2021a) A catalogue of the Thylacine captured on film. *Australian Zoologist* **41**, 143–178. doi:10.7882/AZ.2020.032

Sleightholme SR, Campbell CR (2021b) A catalogue of the motion picture films of the Thylacine (Thylacinus cynocephalus). *Australian Zoologist* **41**, 778–793. doi:10.7882/AZ.2021.026

Sleightholme SR, Robovsky J, Vohralik V (2012) Description of four newly discovered Thylacine pouch young and a comparison with Boardman (1945). *Australian Zoologist* **36**, 232–238. doi:10.7882/AZ.2012.027

Sleightholme SR, Campbell RC, Gordon T J (2020) The Kaine capture: questioning the history of the last Thylacine in captivity. *Australian Zoologist* **41**, 1–11. doi:10.7882/AZ.2019.032

Smith J (no date) Notes held by the Smith family, Launceston.

Smith J (unpublished) Letter to James Fenton NS234/2/1/15. Libraries Tasmania, Hobart, 14 November 1890.

Smith M (1982) Review of the thylacine (Marsupialia, Thylacinidae). In *Carnivorous Marsupials.* (Ed. M Archer) pp. 237–253. Surrey Beatty, Sydney.

Smith S (1981) *The Tasmanian Tiger, 1980: A Report on an Investigation of the Current Status of Thylacine Thylacinus cynocephalus, Funded by the World Wildlife Fund Australia.* National Parks and Wildlife Service, Hobart.

Soriano-Redondo A, Bearhop S, Lock L, Votier SC, Hilton GM (2017) Internet-based monitoring of public perception of conservation. *Biological Conservation* **206**, 304–309. doi:10.1016/j.biocon.2016.11.031

Sprent JFA (1971) A new genus and species of ascaridoid nematode from the marsupial wolf (*Thylacinus cynocephalus*). *Parasitology* **63**, 37–43. doi:10.1017/S003118200006738X

Sprent JFA (1972) *Cotylascaris thylacini*: a synonym of *Ascaridia columbae. Parasitology* **64**, 331–332. doi:10.1017/S0031182000029759

Stannard HJ, Miller RD, Old JM (2020) Marsupial and monotreme milk: a review of its nutrients and immune properties. *PeerJ* **8**, e9335. doi:10.7717/peerj.9335

Stevens B (1988) *Akngwelye atherrerle uthnerreke akete a yeye.* Yipirinya School Council, Alice Springs.

Stoneking M (2018) Mitochondrial DNA. In *The International Encyclopedia of Biological Anthropology.* (Eds W Trevathan, M Cartmill, DL Dufour, CS Larsen, DH O'Rourke, K Rosenberg and KB Strier) pp. 1–3. Wiley, Hoboken, New Jersey.

Stuart D (2019) *The thylacine who guarded Mparntwe*, video recording, ABC Education, broadcast 17 June.

Sugden S, Sanderson D, Ford K, Stein LY, St. Clair CC (2020) An altered microbiome in urban coyotes mediates relationships between anthropogenic diet and poor health. *Scientific Reports* **10**, 22207. doi:10.1038/s41598-020-78891-1

Summers MA (unpublished) Letter to the Commissioner of Police AA612/1/59, Libraries Tasmania, Hobart, 14 May 1937.

Sutton A, Mountain M-J, Aplin K, Bulmer S, Denham T (2009) Archaeolozoological records for the highlands of New Guinea: a review of current evidence. *Australian Archaeology* **69**, 41–60. doi:10.1080/03122417.2009.11681900

Swart S (2018) Resurrection conservation: the return of the extinct? In *Nature Conservation in Southern Africa.* (Eds J-B Gewald, M Spierenburg and H Wels) pp. 130–164. Brill, Leiden, The Netherlands.

Taberlet P, Coissac E, Hajibabaei M, Rieseberg LH (2012) Environmental DNA. *Molecular Ecology* **21**, 1789–1793. doi:10.1111/j.1365-294X.2012.05542.x

Taçon PSC, Brennan W, Lamilami R (2011) Changing perspectives in Australian archaeology, part XI. Rare and curious thylacine depictions from Wollemi National Park, New South Wales and Arnhem Land, Northern Territory. *Technical Reports of the Australian Museum* **23**, 165–174.

Tasmanian News (1888) House of Assembly, Hobart, 27 September.

Taylor L (1996) *Seeing the Inside: Bark Painting in Western Arnhem Land.* Clarendon Press, Oxford, UK.

Temminck, CJ (1824–1827) *Monographies de Mammalogie, ou description de quelques genres de mammifères dont les espèces ont été observées dans lens différens musées de l'Europe. Ouvrage accompagné de planches d'Ostéologie, pouvant servir de suite et de complément aux notices sur les animaux vivans, publiées par M. le Baron G. Cuvier, dans ses recherches sur les ossemens fossiles.* G. Dufour et E. D'Ocagne, Paris.

The Age (1863) The Victorian Acclimatisation Society. Melbourne, 26 November.

The Age (1884) Zoological and Acclimatisation Society of Victoria. Melbourne, 6 February.

The Age (1925) Australian Fauna. Melbourne, 13 August.

The Argus (1884) Zoological Society. Melbourne, 5 February.

The Argus (1950) Tigers off in Tassie. Melbourne, 23 March.

The Argus (1957) And now Copter finds Tasmania's "lost" tiger. Melbourne, 4 January.
The Australian (1826) Native Tyger or Hyena. Sydney, 22 November.
The Canberra Times (1986) Walkers took photos of Tasmanian tigers. Canberra, 9 February.
The Daily Telegraph (1924) Sydney, 19 December.
The Daily Telegraph (1929) New Noah's Ark. Sydney, 17 December.
The Examiner (1907) St. Paul's District. Launceston, 18 January.
The Hobart Town Courier (1829a) Hobart, 14 March.
The Hobart Town Courier (1829b) Hobart, 2 May.
The Hobart Town Courier (1834) Hobart, 6 June.
The Hobart Town Gazette and Southern Reporter (1817) Accident. Hobart, 14 June.
The Hobart Town Gazette and Southern Reporter (1819a) Hobart, 13 February.
The Hobart Town Gazette and Southern Reporter (1819b) Hobart, 24 July.
The Mercury (1921) Personal. Hobart, 28 November.
The Mercury (1922a) Beaumaris Zoo. Hobart, 6 December.
The Mercury (1922b) House of Assembly. Hobart, 10 February.
The Mercury (1922c) The Tasmanian tigers. Hobart, 28 March.
The Mercury (1923) Memorial to the Late Mrs H.L. Roberts. Hobart, 7 August.
The Mercury (1924a) Notes of the Day. Hobart, 26 February.
The Mercury (1924b) Beaumaris Zoo. Hobart, 5 August.
The Mercury (1925) The Game Season. Hobart, 2 April.
The Mercury (1926) Additions to Museum. Hobart, 27 October.
The Mercury (1927a) Tasmanian Tigers. Hobart, 25 May.
The Mercury (1927b) New Arrivals at the Zoo. Hobart, 26 May.
The Mercury (1928a) Beaumaris Zoo. Hobart, 21 February.
The Mercury (1928b) Personal. Hobart, 18 April.
The Mercury (1929) Kangaroo Pest. Hobart, 15 May.
The Mercury (1930a) World's oldest animal. Hobart, 17 October.
The Mercury (1930b) Personal. Hobart, 24 December.
The Mercury (1933) Natural history. Hobart, 27 January.
The Mercury (1937a) Are Tasmanian tigers extinct. Hobart, 10 February.
The Mercury (1937b) Council decides to offer £40 protest at cruelty. Hobart, 9 March.
The Mercury (1937c) Tasmanian tigers. Hobart, 17 February.
The Mercury (1937d) Tasmanian tigers. Hobart, 28 August
The Mercury (1937e) Biological survey valuable. Hobart, 17 November.
The Mercury (1938) Tasmanian Tiger: Traces Seen by Survey Party. Hobart, 5 March.
The Mercury (1947) Tasmanian Tiger Seen by Two Bushmen. Hobart, 1 April.
The Mercury (1948) Tiger's Tracks Seen; Devil Trapped. Hobart, 2 December.
The Mercury (1949a) Huon Resident Claims to See Tasmanian Tiger. Hobart, 18 February.
The Mercury (1949b) "Strange Animal Making Queer Noises" Eludes Searchers. Hobart, 23 March.
The Mercury (1949c) Further Evidence of "Phantom" Animal. Hobart, 4 April.
The Mercury (1949d) "Phantom Animal" in Huon Seen Again. Hobart, 26 May.
The Mercury (1949e) "Phantom" Animal May be Tasmanian Tiger. Hobart, 1 June.
The Mercury (1950) Hobart, 11 February.
The Mercury (1953a) Tasmanian Tiger Caught in Trap. Hobart, 28 March.
The Mercury (1953b) Tasmanian Tiger Only Tiger Cat. Hobart, 30 March.
The Mercury (1964) Hideout of tigers known? Hobart, 19 September.
The Mercury (1978) Tiger cub in 1931 home movie. Hobart, 13 July.
The North West Post (1888) Devonport, Tasmania, 5 May.
The Sydney Morning Herald (1842) Fossil bones. Sydney, 8 October.

The Sydney Morning Herald (1910) A Woman's Zoological Garden. Sydney, 20 April.

The Telegraph (1927) The Melbourne Zoo. Brisbane, 21 September.

The Tasmanian Tiger (1996) [Motion picture]. Winning Post Productions, Hobart.

The Van Diemen's Land Gazette and General Advertiser (1814) Caution. Hobart, 10 September.

Thylacine (2020a) Far Cry Wiki, FANDOM Games Community. <https://farcry.fandom.com/wiki/Thylacine>.

Thylacine (2020b) Zoo Tycoon Wiki, FANDOM Games Community. <https://zootycoon.fandom.com/wiki/Thylacine>.

Travouillon KJ, Phillips MWJ (2018) Total evidence analysis of the phylogenetic relationships of bandicoots and bilbies (Marsupialia: Peramelemorphia): reassessment of two species and description of a new species. *Zootaxa* **4378**, 224–256. doi:10.11646/zootaxa.4378.2.3

Travouillon KJ, Simoes BF, Miguez RP, Brace S, Brewer P, *et al.* (2019) Hidden in plain sight: reassessment of the pig-footed bandicoot, *Chaeropus ecaudatus* (Peramelemorphia, Chaeropodidae), with a description of a new species from central Australia, and use of the fossil record to trace its past distribution. *Zootaxa* **4566**, 1–69. doi:10.11646/zootaxa.4566.1.1

Tucker BW (unpublished) A contribution to the anatomy of *Thylacinus*, W. Tucker archive collection, held by Oxford University Museum of Natural History, Oxford.

Tunbridge D (1991) *The Story of the Flinders Ranges Mammals.* Angus and Robertson, Sydney.

Tyndale-Biscoe H (2005) *The Life of Marsupials.* CSIRO Publishing, Melbourne.

Van Deusen HM (1963) First New Guinea record of *Thylacinus. Journal of Mammology* **44**, 279–280. doi:10.2307/1377473

van Dooren T, Rose DB (2017) Keeping faith with the dead: mourning and de-extinction. *Australian Zoologist* **38**, 375–378. doi:10.7882/AZ.2014.048

Van Dyck S, Strahan R (Eds) (2008) *The Mammals of Australia.* 3rd edn. Reed New Holland, Sydney.

Van Zoelen JD, Kerr I, Prideaux GJ (2020) Description and Identification of Large Mammals Specimens from Nombe Rock Shelter. Unpublished report from Finders Palaeontology, Flinders University, Adelaide.

Veth PM (1993) The Aboriginal occupation of the Montebello Islands, northwest Australia. *Australian Aborignal Studies* **2**, 39–50.

Veth PM, Aplin K, Wallis L, Manne T, Pulsford T, *et al.* (2007) The archaeology of the Montebello Islands, north-west Australia: late Quaternary foragers on an arid coastline. *British Archaeological Reports International Series* **1668**, 1–84.

Vidal N, Marin J, Sassi J, Battistuzzi FU, Donnellan S, *et al.* (2012) Molecular evidence for an Asian origin of monitor lizards followed by Tertiary dispersals to Africa and Australasia. *Biology Letters* **8**, 853–855. doi:10.1098/rsbl.2012.0460

Vilà C, Savolainen P, Maldonado JE, Amorim IR, Rice JE, *et al.* (1997) Multiple and ancient origins of the domestic dog. *Science* **276**, 1687–1689. doi:10.1126/science.276.5319.1687

Wadley JJ, Austin JJ, Fordham DA (2013) Rapid species identification of eight sympatric northern Australian macropods from faecal-pellet DNA. *Wildlife Research* **40**, 241. doi:10.1071/WR13005

Wait LF, Peck S, Fox S, Power ML (2017) A review of parasites in the Tasmanian devil (*Sarcophilus harrisii*). *Biodiversity Conservation* **26**, 509–526. doi:10.1007/s10531-016-1256-x

Waller NL, Gynther IC, Freeman AB, Lavery TH, Leung LK-P (2017) The Bramble Cay melomys *Melomys rubicola* (Rodentia: Muridae): a first mammalian extinction caused by human-induced climate change? *Wildlife Research* **44**, 9–21. doi:10.1071/WR16157

Warburton NM, Dawson RS (2015) Musculoskeletal anatomy and adaptations. In *Marsupials and Monotremes: Nature's Enigmatic Mammals.* (Eds A Klieve, L Hogan, S Johnston and P Murray) pp. 53–84. Nova Science Publishers, New York.

Warburton NM, Travouillon K, Camens AB (2019) Skeletal atlas of the Thylacine (*Thylacinus cynocephalus*). *Palaeontologia Electronica* **22**, 1–56. doi:10.26879/947

Waterworth (1937) Protest on grounds of uselessness and cruelty. *The Mercury*, Hobart, 11 March.

Welch D (2015) Thy Thylacoleo is a thylacine. *Australian Archaeology* **80**, 40–47. doi:10.1080/03122417.2015.11682043

Wells K, Hamede RK, Jones ME, Hohenlohe PA, Storfer A, *et al.* (2019) Individual and temporal variation in pathogen load predicts long-term impacts of an emerging infectious disease. *Ecology* **100**, e02613. doi:10.1002/ecy.2613

Wemmer C (2002) Opportunities lost: zoos and the marsupial that tried to be a wolf. *Zoo Biology* **21**, 1–4. doi:10.1002/zoo.10023

Werdelin L (1986) Comparison of skull shape in marsupial and placental carnivores. *Australian Journal of Zoology* **34**, 109–117. doi:10.1071/ZO9860109

West KM, Stat M, Harvey ES, Skepper CL, DiBattista, JD, *et al.* (2020) eDNA metabarcoding survey reveals fine-scale coral reef community variation across a remote, tropical island ecosystem. *Molecular Ecology* **29**, 1069–1086. doi:10.1111/mec.15382

White JP (1972) *Ol Tumbuna: Archaeological Excavations in the Eastern Central Highlands, Highlands, Papua New Guinea.* Terra Australis 2. Department of Prehistory, Research School of Pacific Studies, Australian National University, Canberra.

White LC, Mitchell KJ, Austin JJ (2018a) Ancient mitochondrial genomes reveal the demographic history and phylogeography of the extinct, enigmatic thylacine (*Thylacinus cynocephalus*). *Journal of Biogeography* **45**, 1–13. doi:10.1111/jbi.13101

White LC, Saltre F, Bradshaw CJA, Austin JJ (2018b) High-quality fossil dates support a synchronous, Late Holocene extinction of devils and thylacines in mainland Australia. *Biology Letters* **14**, 20170642. doi:10.1098/rsbl.2017.0642

Whitley GP (1970) *Early History of Australian Zoology*. Royal Zoological Society of New South Wales, Sydney.

Wilson GP (2013) Mammals across the K/Pg boundary in northeastern Montana, U.S.A.: dental morphology and body-size patterns reveal extinction selectivity and immigrant-fueled ecospace filling. *Paleobiology* **39**, 429–469. doi:10.1666/12041

Wilson J (unpublished) Letter to VDL Co agent Janes Smith VDL22/1/17, Libraries Tasmania, Hobart, 8 April 1879.

Wisbey C (1994) Tiger tales [Radio broadcast], Australian Broadcasting Corporation archive, Hobart.

Woinarski JCZ, Burbidge AA, Harrison PL (2014) *The Action Plan for Australian Mammals 2012.* CSIRO Publishing, Melbourne.

Woinarski JCZ, Garnett ST, Legge SM, Lindenmayer DB (2017) The contribution of policy, law, management, research, and advocacy failings to the recent extinctions of three Australian vertebrate species. *Conservation Biology* **31**, 13–23. doi:10.1111/cobi.12852

Woinarski JCZ, Braby MF, Burbidge AA, Coates D, Garnett ST, *et al.* (2019) Reading the black book: the number, timing, distribution and causes of listed extinctions in Australia. *Biological Conservation* **239**, 108261. doi:10.1016/j.biocon.2019.108261

Woodburne MO (1967) The Alcoota Fauna, central Australia: an integrated palaeontological and geological study. *Bulletin of the Bureau of Mineral Resources Geology and Geophysics, Australia* **87**, 1–187.

Woodford L (2015) The earliest known camera trapping in Australia: a record from Victoria. *The Victorian Naturalist* **132**, 171–176.

Wood-Jones F (unpublished) British Museum (Natural History), London, May 1925.

Woods GM, Howson LJ, Brown GK, Tovar C, Kreiss A, *et al.* (2015) Immunology of a transmissible cancer spreading among Tasmanian devils. *The Journal of Immunology* **195**, 23–29. doi:10.4049/jimmunol.1500131

Woods GM, Fox S, Flies AS, Tovar CD, Jones M, *et al.* (2018) Two decades of the impact of Tasmanian devil facial tumor disease. *Integrative and Comparative Biology* **58**, 1043–1054. doi:10.1093/icb/icy118

World (1924) Round the Ground. Hobart, 6 February.

Worthy TH, Holdaway R (1996) Taphonomy of two Holocene microvertebrate deposits, Takaka Hill, Nelson, New Zealand, and identification of the avian predator responsible. *Historical Biology* **12**, 1–24. doi:10.1080/08912969609386551

Wroe S (2001) *Maximucinus muirheadae*, gen. et sp. nov. (Thylacinidae: Marsupialia), from the Miocene of Riversleigh, north-western Queensland, with estimates of body weights for fossil thylacinids. *Australian Journal of Zoology* **49**, 603–614. doi:10.1071/ZO01044

Wroe S, Clausen P, McHenry C, Moreno K, Cunningham E (2007) Computer simulation of feeding behaviour in the thylacine and dingo as a novel test for convergence and niche overlap. *Proceedings of the Royal Society B: Biological Sciences* **274**, 2819–2828. doi:10.1098/rspb.2007.0906

Wroe S, Lowry MB, Anton M (2008) How to build a mammalian super-predator. *Zoology* **111**, 196–203. doi:10.1016/j.zool.2007.07.008

Wroe S, McHenry C, Thomason J (2005) Bite club: comparative bite force in big biting mammals and the prediction of predatory behaviour in fossil taxa. *Proceedings of the Royal Society B: Biological Sciences* **272**, 619–625. doi:10.1098/rspb.2004.2986

Wroe S, Milne N (2007) Convergence and remarkably consistent constraint in the evolution of carnivore skull shape. *Evolution* **61**, 1251–1260. doi: 10.1111/j.1558-5646.2007.00101.x

Wyuno (1884) Melbourne Gleanings: The Zoological Gardens. *Launceston Examiner*, Launceston, 26 April.

Yalom ID (1989) *Love's Executioner and Other Tales of Psychotherapy*. Basic Books, New York.

Yates AM (2014) New craniodental remains of *Thylacinus potens* (Dasyuromorphia: Thylacinidae), a carnivorous marsupial from the late Miocene Alcoota Local Fauna of central Australia. *PeerJ* **2**, e547. doi:10.7717/peerj.547

Yates AM (2015) *Thylacinus* (Marsupialia: Thylacinidae) from the Mio-Pliocene boundary and the diversity of Late Neogene thylacinids in Australia. *PeerJ* **3**, e931. doi:10.7717/peerj.931

Zhou Y, Wang S-R, Ma J-Z (2017) Comprehensive species set revealing the phylogeny and biogeography of Feliformia (Mammalia, Carnivora) based on mitochondrial DNA. *PLoS ONE* **12**, e0174902. doi:10.1371/journal.pone.0174902

Zieger M, Springer S (2020) Thylacine and Tasmanian devil: between hope and reality–a lesson to be learnt from Google Trends search data. *Australian Journal of Zoology* **67**, 221–225. doi:10.1071/ZO20073

Zieger M, Springer S (2021) Can Google Trends data confirm the need for charismatic species to generate interest in conservation? *Pacific Conservation Biology* **27**, 296–298. doi:10.1071/PC20090

Index

www.ingramcontent.com/pod-product-compliance
Lightning Source LLC
LaVergne TN
LVHW060630110826
845147LV00014B/883